Writing Your
First Play

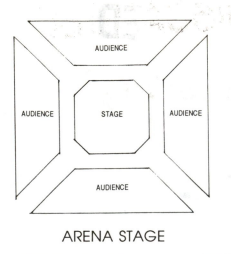

ARENA STAGE

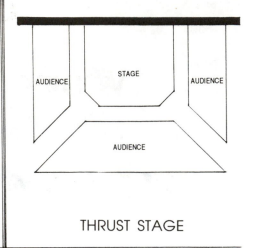

THRUST STAGE

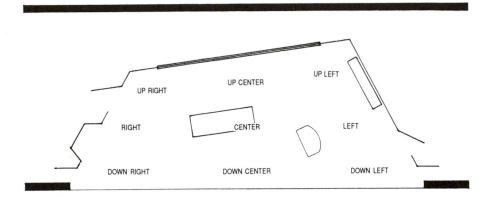

AUDIENCE

PROSCENIUM STAGE

Writing Your First Play

Roger A. Hall

Focal Press
Boston   London

Focal Press is an imprint of Butterworth–Heinemann.

 Recognizing the importance of preserving what has been written, it is the policy of Butterworth–Heinemann to have the books it publishes printed on acid-free paper, and we exert our best efforts to that end.

Frontispiece by R. H. Roggenkamp, Jr.
Cover illustration adapted from frontispiece.

Library of Congress Cataloging-in-Publication Data

Hall, Roger A.
  Writing your first play / Roger A. Hall.
    p.   cm.
  Includes bibliographical references and index.
  ISBN 0-240-80117-2 (alk. paper)
  1. Playwriting. I. Title.
PN1661.H28   1991
808.2—dc20   90-19628
                                       CIP

British Library Cataloguing in Publication Data
Hall, Roger A.
  Writing your first play.
  1. Drama. Composition
  I. Title
  808.2

  ISBN 0-240-80117-2

Butterworth–Heinemann
80 Montvale Avenue
Stoneham, MA 02180

10 9 8 7 6 5 4 3 2 1

Printed in the United States of America

Contents

· · · · · · · ·
Foreword

Phoef Sutton

Writing a play is a snap. All you have to do is arrange a bunch of words in order so that, when actors say them on stage, they will bring an audience to laughter, tears, or enlightenment. Go to it. I'll catch you later.

Okay, so that's an exaggeration, but there are a slew of writing books out nowadays that do deliver the message that the important thing is to *write* the play (or novel, or screenplay) and get that sense of accomplishment. Figuring out how to do so will come to you along the way.

The fact that this theory isn't applied to other professions ("The important thing is to *do* the heart/lung transplant. You'll figure it out as you go.") doesn't seem to bother anyone. Plays are emotional things; things from the heart. They aren't something you have to learn how to do.

Or do you? Faith and inspiration might get you a good scene or even a good act, but eventually you'll see that those flashes of inspiration need to be made part of a structure, that a play has to be built. In order to build something, you need to have material and tools. That's what this book provides.

I've been working as a writer for ten years in theater, film, and television, and I've only taken one writing course. It was the first one I ever took and it was taught by Roger Hall. I hadn't written before then. I haven't stopped since.

I remember one exercise in particular (it's in this book) that changed my whole idea of what a play could be. Open a script and the eye clearly sees what a play is—people standing around talking, sometimes sitting around talking. So when Dr. Hall assigned Exercise 1—write a scene with no dialogue—it seemed like a weird, if fun, stunt. As I wrote the scene (as I recall it was about something startlingly original, like people in dorms playing loud music), I saw that it didn't have to be strained, or forced, or twisted to fulfill the requirements. (If it was all three, that wasn't the assignment's fault.) I saw that a scene could be complete and whole without words, because a play isn't about words; it's about people doing things, and speaking is only one of the things they do.

It sounds obvious. It's a simple lesson, but of the hundreds of scripts I've read since I became a producer, 99 percent of them have been nothing but people talking.

All of the exercises from Dr. Hall's class, each a giant leap forward in knowledge and experience for me, are in your hand right now. Also in it, and most entertainingly, are my fellow students. One of the most delightful aspects of this book are the writing samples from Dr. Hall's students over the years.

Most writing books, in their section on conflict, ask you to read, say, a scene from *Glengarry Glen Ross,* which leads any sensitive novice writer to give up and go into frozen produce marketing. You won't be able to write like Mamet right off the bat and forcing a comparison like that is recipe for writer's block. If you're serious about being a writer, you've read the great and the good plays; you know what to aspire to. Now you're trying to learn your craft.

That's a lonely process, and the other students in this book will keep you company, give you someone to compete with. Someone to make you say, "I can do that," "I can do better than that." (Though you may be surprised; some of this writing is quite good.)

So do it. Read the book, do the exercises. Have fun, talk back. When you're through, you'll be ready to tackle that heart/lung transplant without fear of losing the audience on the operating table.

Phoef Sutton is executive producer and writer for the NBC television series "Cheers!" for which he won an Emmy Award in 1989. He has been a recipient of a National Endowment for the Arts Playwriting Fellowship, and he was the winner of the Norman Lear Award for Comedy Playwriting and the Roberts/ Shiras Playwriting Award.

Preface

As a student, I took a playwriting class in which the teacher, on the first day, said, "Your first assignment is to write a play. Bring it to class in two weeks and we'll read and discuss it."

After class that day I turned to the fellow student beside me and asked, "Do you know how to write a play?"

"No," she replied, "that's why I took the class. Do you?"

No. I didn't either. I realized intuitively that there had to be a better way to help people write plays. When, as a teacher myself, I began teaching playwriting, I put my efforts together with Dr. Ralph Cohen, a colleague in the English Department, who had similar interests.

We wanted to devise an approach that would allow students to work on certain fundamental aspects of playwriting one element at a time. We also wanted to encourage plays to evolve from one idea to another.

After some trials and errors, we developed a series of exercises that helped students to understand and use basic elements of drama such as action, conflict, dialogue, and character.

It just so happened that our experiments with playwriting came at a good time, for the last few years have seen a burgeoning interest in playwriting. Playwriting contests have sprung up like wildflowers, sometimes in the most unlikely terrain, and many of them carry substantial financial and artistic rewards. Community theaters, secondary school drama programs, and college theater departments have demonstrated a greater willingness than previously to depart from the Broadway-hit syndrome and experiment with original works. The same is true of regional professional theaters, many of which have become prominent in the development of new scripts. In addition, almost every major city has semiprofessional or professional companies that focus their attention, sometimes exclusively, on original drama.

Broadway—the New York commercial theater—which in many ways is still regarded as the pinnacle of American theater, has also responded to this surge of new plays. Because of high costs, most of the productions in the major New York theaters have hued to the economically safer road of revivals or imports of successful British plays. Even in New York, however, such original American plays as *Steel Magnolias* and *Driving Miss Daisy* have flourished. In many cases scripts such as *Fences*, *The Gin Game*, and *The Grapes of Wrath*, which were developed and produced first in other areas of the country, have eventually enjoyed Broadway success. Also numerous off- and off-off-Broadway theaters in New York have gained a certain amount of notoriety through their production of new material.

This current fascination with the production of original scripts has led to explorations of the best and most effective ways to put a story into dramatic form. That is hardly a new concern. Ever since Aristotle some 2,300 years ago tried to label the elements that comprised a superior tragedy, writers and critics have been attempting to tell people how to construct plays. Writers of our own century have not been idle in that enterprise. Dozens of volumes have been printed on playwriting, playwrighting, playmaking, how to write a play, and even how *not* to write a play. They have been written by knowledgeable critics of the drama such as Walter Kerr and Brander Matthews, experienced teachers of playwriting such as George Pierce Baker and Sam Smiley, and successful playwrights such as John Van Druten (*I Remember Mama* and *I Am a Camera*) and John Howard Lawson (*Processional*).

Every one of those dozens of volumes about playwriting contains valuable nuggets of advice—many of them reappear in the chapters of this work—and many of the books are excellent guides for the analysis of plays and the understanding of the playwright's craft. Still, as a teacher of playwriting, I thought something was missing.

The development of a play is too often viewed in terms of procedure: Proceed from an idea to a brief scenario to an expanded scenario or outline to a rough draft of the play. Sometimes the development of a play is seen in terms of a progression of analytical elements: Progress from an inciting incident through a series of minor conflicts or crises to a major crisis (the climax) and thus to the conclusion or denouement. Both of those standard methods seem to assume that the writer has the ability to invent a relatively complete idea for a play and needs only some working procedures or the inclusion of various standard dramatic devices in order to bring the idea to fruition.

My teaching experience suggested otherwise. I saw that young men and women who had never before written a play had much more basic needs. First, they needed some experience in working with the fundamental building blocks that make a play a play—such things as action, conflict, and the interaction of characters. Second, they needed help in generating an idea for a play. The concept of "getting an idea" for a play is a very complicated notion. Ideas for plays seldom emerge fully grown from the playwright's mind. More frequently they take root as a seedling. They grow and develop bit by bit, with the introduction of a scene here, a new character there, and an idea or two over there. Eventually, the idea progresses and the sapling takes on the stature and the shape of a mature tree with trunk, limbs, and branches. The foliage is brilliant with vivid characters, dramatic action, and snappy dialogue.

Just as it is hard to imagine how a mature tree will grow from a seedling, it is extremely difficult for a beginning writer to envision a fully developed play. It is also almost impossible for an inexperienced writer to manage with any degree of skill the many essentials of a play—from action, conflict, and environment to dialogue and characters—all at one time. It would be as if a music teacher were to tell a young student who was learning composition, "I want you to compose a symphony," without first instructing the student in harmony, progression, syncopation, and the other necessary musical elements. Complex creations such as symphonies, plays, and trees unfold in stages, and those who are learning to create them need to practice with one or only a few elements at a time before combining them. The approach I suggest allows plays to develop. Through the sequence of exercises I describe, scenes become progressively longer, more complex, and more sophisticated until, like leaves in spring, plays appear.

This book is designed to help the beginning and inexperienced playwright. It presents for students a series of exercises to provide practice in working with some of the essential ingredients of a play. Because the exercises are sequential, one leads naturally to another, becoming more complete and more complex with each step. The exercises provide an opportunity to begin with a seedling of an idea and nurture it carefully before asking it to bear the fruit of a mature tree.

This approach has been quite successful with the undergraduate students I teach. In the dozen years that I've been working with original scripts, James Madison University has produced more than 50 student-written plays, including one-acts, full-length pieces, and musicals. Most of those were mounted in our experimental theater, but several were a part of our main season of plays. Six of the plays were recommended for competition by the American College Theater Festival, and five received awards in various writing competitions.

Some of the students who have used these exercises are still writing successfully for the stage, for the screen, and for television. Many have gone into professional theater in other capacities, as managers or performers. Several are writing for newspapers or magazines, and others are teachers, including some who are using these exercises with *their* students.

In addition to the work I've done with my own students, I have also shared these exercises with English and drama teachers at the high school level, and they have reported using them successfully with their students. The exercises also formed the basis of a playwriting unit for high school students at a young writers' workshop at the University of Virginia.

In this book I make numerous references to a variety of plays and movies—classic and modern—from Sophocles and Shakespeare to Arthur Miller and David Mamet. In many cases, however, I've used hypothetical examples or short scenes written by students. I've done that for several reasons. Although great writers can provide marvelous models, their works are often longer and more complex than I want for the illustration of particular points. Also, they do not reflect what someone using these exercises might actually write.

The student writings, on the other hand, provide concise examples of exactly what the exercises can be expected to produce. They are also relatively short, and they are finished. That is, even if they suggest a continuation, they represent a whole response to an exercise rather than a scene taken from a longer play. For that reason they are simpler to work with than a scene that is part of a more complex project.

Many people contributed to this book. I would like especially to thank Dr. Ralph Alan Cohen of the English department who worked with me to originate this sequential approach to playwriting. I have appreciated the assistance of James Madison University for this project, and in particular I want to express my gratitude to my colleagues in the theatre program who have encouraged the efforts of student playwrights. Finally and most importantly I want to thank my playwriting students who have consistently amazed, surprised, and entertained me with their work and who generously permitted me to use examples of that work in this book.

Introduction

The direct and primary goal of this book is to enable a person to write a play of at least one act in length. A one-act play is usually, though not always, a piece of continuous action in one setting, running about 30 to 40 minutes in playing time. That, however, is merely a guideline. Some one-act plays are as short as five or ten minutes. Others use two or more different scenes or units of actions, while still others use locations that shift from one place to another.

The purposes of this book are all relatively simple. If you are interested in writing a play, it provides a series of exercises designed in sequence to help you do that. You should read the assignment, the explanation, the examples, and the evaluations. Then you should do the assignment. Then compare your writing with the examples and look at it in light of the commentary on those examples. The one important part missing, of course, is someone with an artistic bent to provide an outside opinion. If you are using this guide on your own, you will have to provide your own objectivity.

Another purpose is to enable a teacher to provide a basic course or unit in playwriting. Most teachers of theater have the intelligence and artistic sensibility to provide helpful comments on inexperienced writers' first attempts at playwriting. Many, however, are unsure about what assignments to use in the actual development of a play. The step-by-step orientation of this book is designed to solve that problem.

Not everyone who sets out to write a great play will be successful, but everyone who attempts the exercises here will come away with three very important, indirect benefits. First, a writer experimenting with the dramatic format will gain a greater appreciation of plays, of the ways in which they are constructed, and of the choices the playwright makes. Second, the exposure to working in a dramatic mode will help a writer in other formats. The emphasis on action, character, conflict, and dialogue will affect the way a writer thinks, even the writer who is later working in poetry or prose. Third, the writer will gain a greater awareness of human behavior, of the intricacies of personal relationships, of the reasons why people do things to each other, and of the ways they react to their environments.

People reacting to their environments: That's one of the primary concerns of drama, and it's one of the reasons why the first section of this book deals with "action."

1

Action

EXERCISE 1 • DESCRIBE AN ACTION
Write a description of an action taking place. Describe only those things that can be seen. Use no dialogue, although you may use other sounds. The scene should take place in one location.

Drama is the imitation of an action. Unfortunately, most of us experience plays by reading them rather than seeing them. It seems hard to believe, but there exist countless individuals who have never seen a live play performed. Even for those of us who have seen scores of plays, we've probably read more than we've seen.

When we read a play, we read the characters' names and the lines they speak. Very little of what the characters *do* is described. That's particularly true of classical plays. Shakespeare's plays, for example, which are often an individual's first exposure to drama, have little in the way of stage direction or descriptions of particular actions. Rather, in those plays, most of the action must be deduced from the spoken line.

When Macbeth returns from killing King Duncan, only a reading of one line of dialogue three quarters of the way into the scene informs us that Macbeth has unthinkingly brought the murder weapons with him instead of leaving them behind. In performance, of course, the action would provide that information immediately. When Macbeth enters, we would see the knives in his hand, and we would continue to see them there throughout the scene.

The fact that most of us receive our introduction to plays through print rather than performance has serious consequences for beginning playwrights. Since, for the most part, we only read the lines spoken by characters, is it any wonder that we come to think of plays as a dialogue between characters in which the words are all-important and tell us everything we need to know? And is it any wonder that when beginners sit down to compose a play, what usually develops is a group of characters sitting around talking?

Our first task, then, is to create a different perspective on what constitutes a play. We must find a way to present drama in terms of actions rather than words, in terms of what people *do* rather than what they *say*.

Action is one of the building blocks of all drama. Drama, after all, is about what people *do,* and what people *do* is *action.* Hence it is instructive to look at the idea of action in some detail.

In brief, action occurs when someone does something. But the concept of action is really much more complex and multilayered. Let's look at an example: A woman walks into a fast-food restaurant. She stops and looks up at the items available and their prices. She gets into line. She pulls her wallet out of her purse and takes money out of the wallet. She reaches the front of the line and requests "A cheeseburger, small fries, and a small orange." She pays for the food, takes it, walks to a table, and begins to eat her meal.

There are numerous actions taking place in this brief scene. Just walking into the restaurant is one action. Reading the menu is a second. If the woman puts on a pair of glasses while she's reading, that's a third. And so on. It quickly becomes clear that not all her actions are equally important. Walking in is probably not as important as ordering the food, or as eating it. Hence, even the simplest actions have a hierarchy of importance with respect to other actions. It is crucial for playwrights as well as directors and actors to understand the way in which actions fit together and to decide which actions are the most important.

Remember that not all actions are equally important—that all actions are important in different ways. I suggested that walking into the restaurant was not as important as ordering or eating the food. Nevertheless, walking in is important in that it gets the woman into the restaurant. In that sense it's essential. The rest of the scene can't take place without it. In some instances the entrance of a person might be the most significant action taken. The woman, for instance, might hurry into the restaurant and then look carefully behind her out the window.

Reading the menu is another action. Is it important? In one sense it is not. It could be eliminated and the woman could still get her meal. But the action provides significant information. That the woman looks at the menu might indicate that she doesn't frequent the restaurant, that she's indecisive, that she's concerned about how much things cost, or any number of other pieces of information. Any of those elements might be extremely significant. The scene would certainly be different if the woman simply walked in, walked up to the counter, and ordered.

In both cases, whether she reads the menu first or not, the woman would get her meal. Here is where actions become multilayered. This scene contains many actions. Taken together, however, they constitute a larger action: A woman buys her lunch. If we also saw her eating her meal and leaving, the whole action might be "a woman lunches."

The same layering of actions occurs in any play. In one moment Hamlet praises an actor. That is one action. A series of actions taken together might constitute a larger action: Hamlet sets a trap. All of those actions taken together form the action that dominates the character throughout the play: Hamlet avenges his father's murder. In acting terms that overall action is what is meant by such terms as *through line, superob-*

jective, or *spine.* The recognition that small actions fit together to create larger actions is just as important for a playwright as it is for an actor.

Taken simply, an action is someone doing something. The action can be small in scope or large; it can be a simple action or a complex action composed of numerous smaller actions. It can be static action—such as sitting or sleeping—or frenetic action—such as running or dancing.

Most modern theater practitioners also recognize psychological action—a thought process, a decision, or a point of view. For instance, if a lawyer tells you to do something, his authority or his strength might compel you to do it without even the slightest actual physical action on his part. There is a psychological force at work.

For beginning playwrights, however, I've found it more helpful to consider thought and decision as preceding physical action. Picture, for example, a woman in a department store. She looks at a bracelet. Then she looks around. Thought is occurring. A decision is reached. The woman taps a bell on the counter to call a salesperson.

At the moment that the woman is looking around, we may also think she is considering pocketing the bracelet. The action that follows—tapping the bell—reveals the thought process that precedes it.

At this beginning stage, we want to focus on basic action—such as the woman ordering her lunch—that is large enough to comprise several other actions. It is enough that we understand that the action may in turn be part of a larger, more complex whole.

We must next understand that action by itself, be it simple or complex, static or frenetic, does not guarantee interest. Now for some circular thinking. For an action to gain the interest of the audience, it must be dramatic action. But what is dramatic action? And how is it differentiated from other action? Dramatic action is action that gains the interest of an audience.

William Archer, the British dramatist and critic who translated Henrik Ibsen for English-speaking audiences, wrote nearly the same thing. His turn-of-the-century study of playwriting, which still represents sound thinking, defined "dramatic" as "any representation of imaginary personages which is capable of interesting an average audience assembled in the theater."

The first time I read that I regarded it as totally inadequate. Surely, I thought, we should be able to determine what is dramatic within the context of the script itself. But with his vast experience in theater, Archer realized, as I did only somewhat later, that the audience is essential to the play; that just as an architect's plan must undergo stress tests to determine its strength, so a play must undergo the test of an audience or experienced readers to determine its dramatic content. Dramatic action is realized only when what the author *thinks* is dramatic is effectively conveyed to others.

George Pierce Baker, a famous teacher of playwriting who helped to develop the talents of Eugene O'Neill and Philip Barry, realized that, too. "A play," he wrote, "exists to create emotional response in an audience."

A perceptive playwright can gather some notion of what is likely to be dramatic activity. To explore this concept, let's look at dramatic action from several different angles. One approach is to see dramatic action as action based on the desire of a human being to attain a goal. A man walks onto a stage with a painting and an easel and begins to set up the easel. It falls over. He tries again. It falls again. He carefully inspects the easel, looking at the legs and hinges. He tries yet again to set it up.

In this instance we understand the goal—to set up the easel. Perhaps we infer the larger action, to display the painting. As the man tries to accomplish his goal, another element of dramatic action comes into play—namely, tension or conflict. The human will to attain the goal is being thwarted by the inanimate object. The man is in conflict with the easel. As a result of the conflict, tension arises.

The human will to attain a goal. Conflict. Tension. All of those elements work together to induce *questions*. Will the man succeed in setting up the easel? What if he doesn't? What will he do then? What does he want to do with the painting?

The scene could continue in a variety of ways. The easel falls again. The man picks it up and bashes it to pieces. Then he picks up the remains of the painting and calmly walks off the stage. An alternative: The easel falls again. The man goes offstage, comes back with a chair, sets that on the stage, and displays the painting on the chair. Another alternative: The easel stays up. The man sets the painting on the easel. In all cases the questions have been answered, and the action is ended.

Ingmar Bergman, the Swedish director and dramatist, once commented on how he formulates ideas. Bergman explained that he literally has visions. He might visualize two women in a pink room. One might be turned away, looking out a window. Bergman would ask himself questions about his vision. Why was the one woman turned away? What was outside the window? Why was the room pink? What was going on between the two women? If the answers to his questions were interesting to him, he'd ask more questions until, perhaps, he had something to write. Essentially, said Bergman, he wrote not because he had something to say, but because he had questions to ask.

Jon Jory, the producing director of Actors Theatre of Louisville and one of the primary forces in the development of new playwrights in the United States, has said that a play should always pose a question within its first five lines. Two or more questions, he noted, would be even better.

But, significantly, the questions need not come in the form of dialogue. In fact, in good drama the questions are usually inherent in the *action*, and the dialogue—questions expressed verbally—simply adds to the questions already raised implicitly by what we see occurring.

The opening scene of Arthur Miller's *Death of a Salesman* illustrates how dramatic action provokes questions. Willy Loman, the sixty-year-old salesman, enters carrying two large sample cases. He crosses to the doorway of his dimly lit house. He unlocks the door, comes into the kitchen, and lets the cases down. He feels the soreness of his palms and a word-sigh escapes his lips. He closes the door and carries the cases into the unseen living room. We see Linda, his wife, stir in her bed. She listens to Willy's entrance, then gets up and puts on a robe. The dialogue begins:

```
Linda: Willy!

Willy: It's all right. I came back.

Linda: Why? What happened? (Slight pause) Did something
       happen, Willy?

Willy: No, nothing happened.

Linda: You didn't smash the car, did you?

Willy (with casual irritation): I said nothing
       happened. Didn't you hear me?

Linda: Don't you feel well?

Willy: I'm tired to the death.*
```

We *see* the action as a tired old man enters his house late at night. We *see* a woman obviously concerned about the man. He wants to rest. She wants to know what's going on. Each person wants to attain certain goals. We *see* the man dismiss her concern. The human wills are in conflict. We *see* a relationship straining. Tension.

This excellent opening scene induces numerous questions. Why is he coming in so late? Why is he tired? Why is she concerned? Those are merely three questions out of many. And all of that has been provoked by the simple action that occurs even before a line has been spoken.

Miller then uses a variety of questions in his first few line to focus our questions. Some of them are significant. "What happened?" asked twice tells us that Willy wouldn't be there if something weren't wrong. The question about the car informs us he has the capability of smashing it, as he does at the end of the play.

There are significant elements that are *not* present in my summary of this opening scene. In my condensation of Miller's opening stage directions I omitted, among other things, the following reference to Willy: ". . . his exhaustion is apparent." I omitted it because I wanted you, the reader, to make that conclusion based on Willy's *actions:* feeling his sore palms and sighing.

Just as Willy's exhaustion is exposed through his actions, so dramatic action is expressed in verbs, in what people *do*, not in moods. It is extremely difficult for an actor to play a general mood such as anger, and it is well nigh impossible for a playwright to write an effective general mood. The dramatic mood arises from the specific actions undertaken.

When we say to ourselves, "That person is angry," or "That man is exhausted," we are making a conclusion based on certain actions we have seen performed. We do this so frequently and so automatically that we hardly even think about the actions we are seeing that prompt us to leap to such conclusions. We just leap.

Test this. The next time someone describes another person as "angry" or "happy" or "sick," find out why that person thinks that. The conversation is likely to run something like this:

```
Friend: Did you see Frank today? He really looked
    awful.

You: Oh yeah? How come?

Friend: Well he didn't look like he felt very good.

You: Why do you say that?

Friend: I don't know. He just looked sick.
```

There it is. One conclusion ("He looked sick") used to justify another conclusion ("He didn't look like he felt very good") to justify another conclusion ("He really looked awful").

Perhaps the most startling comment made by your fictional friend is "I don't know." Of course your friend knows. The friend must know, or else how could your friend have arrived at the conclusion that Frank is sick. Perhaps Frank was moving more slowly than usual, or was hunched over. Perhaps he coughed, sneezed, grimaced, groaned, or limped. All of those are *actions,* and whatever the symptoms, your friend saw those actions. But instead of registering "Frank is coughing and talking hoarsely," your friend automatically registered the conclusion of a general mood: "Frank is sick."

The problem with that normal and generally quite effective way of proceeding to conclusions is that many times the conclusions cannot be seen, and playwrights must deal with what can be seen.

Let's look at another example of the difference between a general mood, which cannot be seen, and a specific action, which can be seen and which establishes the necessary mood. A hypothetical stage direction reads, "Jennifer enters the room. She is angry." That's a problem. Nothing is visible. We have no outward manifestation (action) to suggest the inner emotion. On the other hand, the stage direction might read: "Jennifer hurries into the room. She stops and looks around. Her eyes land on the stereo system, and she moves to it. She quickly looks through several cassette tapes, and then selects one. She yanks the tape out of the cartridge, throws the cartridge to the floor, and smashes it with her foot." That is a dramatic action. We *see* the action. We understand the mood. And we are provoked to ask questions about what is going on. We might wonder not only why Jennifer is so upset but also why that one particular tape was so significant.

This notion of looking for the action that prompts conclusions rather than settling for the conclusion is equally important to performers. Playwrights occasionally slip, and when they write "Jennifer is angry," it's the performer who has to translate the generality into concrete action that is both quintessentially human and spellbinding.

Many individuals turn to writing because they think they have something to say. In writing plays, that can be a problem. I have a rule that I repeat over and over again to beginning playwrights: DON'T SAY IT, SHOW IT!

Thornton Wilder, in his superb essay on playwriting, expressed it this way: "A play is what takes place. A novel is what one person tells us took place." So if you have something to *tell* us, write a novel. If you have something to *show* us, write a play.

With that as a reminder, let's proceed to the first exercise.

● ● ● **EXERCISE 1**

Write a description of an action taking place. It should be no longer than two pages. Describe only those things that can be seen. Use *no dialogue*, although you may use other sounds. The scene should take place in one location, but you do not need to concern yourself with stage terminology.

Let's look at two very basic examples—one acceptable, one unacceptable.

EXAMPLE

It is dark. A man is lurking in the bushes beside a house. He looks around, then slides carefully to a window. He looks around again, then tries to lift the window. It doesn't move. He takes a small tool from his pocket and scores a square on the window. He hits it with a gloved hand, and a section of the window drops out. He reaches his arm inside the window. A light comes on in the room. The man quickly withdraws his arm, turns, and begins to run.

EXAMPLE

It is a dark room. A man and a woman are sleeping. The alarm clock goes off. The man switches off the alarm. It says 6 A.M. The man takes a gun from beneath his pillow, looks at it, then places it beside the clock. The man continues to sit, groggily thinking of the tequila he guzzled the night before and the short rest he has had. He rises and glances at a paper on the table. A headline reads "Bank Heist Sets Record." He goes to the closet where he takes out pants

and a shirt. He begins to put on the shirt
and cannot button the sleeve because a
button is missing. He strides angrily to the
bed, and wakes the woman. He holds out the
buttonless cuff. She shrugs and turns over.
The man finishes putting on the shirt as
best he can. He puts on his pants, finds his
socks and shoes, puts them on and exits.

EVALUATION

As you read the scene, can you picture the action as it occurs? Is anything missing or unclear? Would you, for example, be able to see a headline or a small clock on a bedside table? Probably not on a stage. Even with a movie close-up, you probably still wouldn't know if it were 6 A.M. or 6 P.M.

I indicated that questions are important to dramatic action. Questions raised can also be a problem for a play, especially if there are too many questions or if significant questions are left unanswered. Examine what questions are raised by the action and whether they are answered. In the first example, the key questions are "Will the man succeed in breaking in?" and "Will he be discovered?" Both of those are answered. In the second example we see a man getting dressed and upbraiding the woman in the room. But what about the gun? What's it for? And the bank heist? Did he do it? Who knows. The author is trying to include too much, and the information becomes confusing.

Next ask yourself, "Does the example contain anything that cannot be seen?" The second example has numerous problems in this area. There is the difficulty of the time, for one thing; the newspaper presents a similar problem. But more importantly, we cannot read characters' thoughts except through their actions, so we cannot possibly know that the man is thinking of tequila or his short rest.

The novelist or the short story writer can take us directly into the mind of a character. If the novelist writes, "Turk sat groggily on the bed thinking of the tequila he'd guzzled the night before and the short rest he'd had," then that's what is on Turk's mind.

The playwright does not enjoy that option. Even if a character speaks directly to the audience—as Iago does in explaining his actions in Shakespeare's *Othello*—the character may or may not be telling the truth. Some characters may not even fully understand their own motivations.

The playwright must find a means to make visible to the audience the outward manifestation of the characters' thoughts. For example, the playwright might write: "The man rises unsteadily and begins to walk slowly to the closet. He glances at an empty bottle of tequila lying on its side on a table. He stops, picks up the bottle, and holds it upside down as he watches the last drops fall on the table. He pitches the empty bottle in a trash can and proceeds to the closet." That would clarify the relationship between the man and the bottle.

There are moments in plays when we, as the audience, feel as though we know exactly what the character is thinking. In those moments the work of the playwright and the performer come together in such an effective way that the external activity makes clear the inner thought process. That activity can be as subtle as a look or a shrug or as overt as a statement or a punch. In any case, the playwright and the performer are carrying us along, and we seem to understand why characters are behaving as they do.

Are there places where the scene seems to need words? In the first example there are none, although voices might be heard when the light comes on. In the second example the moment between the man and the woman seems to want dialogue. In that instance lines could add something to the action.

There is another question: "Does the scene suggest a mood?" The mood of the first scene is definitely melodramatic. You can almost hear suspicious-sounding music underscoring it. The mood of the second piece is indistinct. The style is realistic, but what is described could be played comically or despairingly.

A final question about the description is, "What kind of characters are expressed through the actions?" In the first example the man in the bushes seems calm, orderly, and experienced. In the second example the man seems not entirely in control of his situation. The woman in the bed certainly pays no heed to his remonstrations.

Put your skills of evaluation to work on one final example, this one written by an undergraduate student.

EXAMPLE

A MAN IN A BUS TERMINAL
by Lynda Edwards

A tiny middle-aged man wearing a rumpled
suit is sitting in a big-city bus terminal.
It is night. The clock on the wall says
11:45. He has a small overnight bag with
him. The man glances frequently at his watch
and at the clock on the wall, and at two
disheveled bums sleeping on the floor near
the doorway. A muscular man wearing a T-
shirt proclaiming "Olympic Sex Team" under a
leather jacket enters and flops down next to
the man. The man nervously unfolds a
newspaper and tries to read as the Leather
Man glances at the paper over his shoulder.
The paper gets tangled and crumpled as the
tiny man desperately tries to smooth it.
Finally he smiles hesitantly at the Leather
Man, wads the paper into a ball, stuffs it

under his arm, and lunges toward the ticket counter with his coat and bag.

He pulls out his ticket from his pocket and slaps it on the counter. The attendant behind the counter slowly walks over, eating a sweet roll. The small man slaps the ticket angrily and points to the clock behind the counter. The attendant holds the roll in his mouth--crumbs and powdered sugar falling to the counter--as he inspects the ticket. Then he glances at the clock, shrugs, hands the ticket back to the man, and walks away.

The man stuffs the ticket back into his pocket and goes to the bathroom, which is a small, windowless room. He slams the door and turns the lock vigorously. He splashes water on his face, sighs, takes a deep breath. He smiles, looks at himself in the mirror, straightens his shoulders, and reaches for a paper towel. There are none. He shakes his hands as he marches for the door. He cannot get it open. He yanks and tugs. He puts one foot on the door and pulls. He backs up, laughs, and shrugs. Then he whirls and attacks the door, tugging hysterically. He calms himself. He knocks timidly, then louder. He pounds on the door with both fists. He kicks the door. He looks around the bathroom. He goes to the sink and suds up the soap. He splashes the wet liquid on the door knob and lock mechanism. He tries again to open the door, but his hands just slip. He looks at his wet hands. He reaches again for paper towels that aren't there, and then hits the metal container. He dries his hands and the door handle with his coat. He goes to the corner of the bathroom and sits down with his bag. He takes a package of Rolaids out of his bag and pops one into his mouth. He makes a pillow out of his coat and puts it behind his head. He leans back, curls up in the corner, and closes his eyes.

THE END

EVALUATION

This lighthearted piece answers all the questions it raises. The author uses virtually every small detail that she establishes. The man's suit becomes both his towel and his pillow, and the Rolaids come from the overnight bag. Furthermore, everything can be seen. Even the clock in the bus terminal would be large enough, and, if it were dark out and the lights in the terminal were on, it would be apparent that the time was evening.

Although everything in the scene is clear without words, dialogue might add to the enjoyment of the piece in two places. We might enjoy a conversation between the small man and the messy attendant. And wouldn't the small man call out in his attempts to get out of the bathroom?

The important thing to remember is that in those two instances lines could add to the action. Thus we can understand dialogue as being a complement or an accessory to the action. Words, rather than being an end in themselves, become merely one method that people employ to accomplish certain aims or perform certain actions.

The scene is obviously comic in mood, as would be established in the meeting of the small man and the imposing (though apparently harmless) Leather Man. As for character, the scene presents a timid, unfortunate soul not dissimilar from some of the great silent movie heroes. You can almost visualize Charlie Chaplin, Buster Keaton, or Harold Lloyd confronting the Leather Man, the attendant, or the locked door.

As you prepare to write your own "silent scene," you might wonder where the ideas for the scenes come from. They can come from anywhere—from your observations or from your imagination. If you want a starting mechanism, I suggest you think of a person in a particular environment, something as simple as a man on a beach looking around. Then, like Ingmar Bergman, ask yourself questions. What is he looking for? Something he lost? Someone he knows or is waiting for? A wallet? A key? A missing child? Situations and actions immediately arise.

Another prod to the imagination is to try to find something incongruent or out of place in a situation—like one of those "what's wrong with this picture?" puzzles. Suppose that the man on the beach is in a suit and tie, or carrying a common but un-beach-like article such as a briefcase. Again, the questions will lead you to a situation and an action. If they don't, try a different person in a different environment.

Writing an action is a major step for a playwright because it establishes a particular viewpoint toward drama. It demands a mind-set geared toward seeing a play in terms of action, in terms of people doing things to each other and interacting with each other. That concept of drama as action will remain the basis for the following exercises. Therefore, as we move on to more complex exercises that integrate various elements of drama, always keep one question foremost: "What do we see these people doing?"

2

Direct Conflict

EXERCISE 2 • DIRECT CONFLICT
Place two characters in a scene of direct conflict. Write the
scene in stage terms, indicating how the scene should be
staged. Use dialogue as needed.

This chapter asks you to compose a scene that places two characters in a
situation of direct conflict. The exercise encourages you to conceive a dramatic scene in terms of conflict. Taken in combination with the first exercise, it attempts to substitute for the notion of people sitting around
talking the more dramatically functional notion of people doing things
that are in conflict.

The idea that conflict is essential to drama is not a new one. Fer-
dinand Brunetière expressed it at the turn of the century as an extension
of the nineteenth-century dialectical theories of Friedrich Hegel. It has re-
mained a tenet of dramatic criticism ever since. More importantly, the
earliest dramatists have instinctively known that conflict leads to good
drama. Antigone confronts Creon. King Lear's daughters reject him.

The nineteenth and twentieth centuries saw various attempts to
produce drama without traditional conflict. Naturalistic, "slice-of-life"
drama was an attempt to study human beings in a scientific manner, to
examine daily lives without resorting to contrived conflicts. In the world
of the 1990s, music videos and avant garde theatrical presentations often
juxtapose words, sounds, and visual images, which are devoid of traditional conflict, in an effort to induce emotional responses. The twenty-
first century is likely to see more "montage drama," which will seek to
elicit emotional reactions through the manipulation and layering of various aural and visual stimuli.

Despite the emergence of those and other "nonconflict" dramatic
modes, conflict has been and remains such a fundamental ingredient of
storytelling that a beginning playwright needs to confront conflict head
on.

The desires of different characters come into conflict within plays
as a whole and within individual scenes within those plays. Sometimes
the conflict is psychological—for example, when Hamlet considers his
choices or bandies with Polonius; sometimes it is physical—when Hamlet
duels with Laertes. Sometimes the conflict is subtle—when Macbeth cal-

culates with his wife; and sometimes it is overt—when Macbeth finally meets Macduff. Whatever its shape, conflict provides an important starting place for a beginning playwright.

Conflict does not arise out of nothing. Conflict is intrinsically tied to two other components: intention and outcome. For conflict to exist, there must be an intention or a goal. There must be a character who wants something and who undertakes action to achieve that intention. If someone or something prevents or hinders the immediate attainment of that goal, then conflict exists. Eventually some resolution to the conflict occurs, and that is the outcome.

The hindrance—and therefore the conflict—can arise in three ways. It can be internal—a physical or mental element within the character. A boxer who is growing old is in conflict with the internal aging process of his own body as well as with his external opponent. A young artist trying to decide if she should risk making it or settle for a more secure future is battling internal psychological demons.

A second type of conflict or hindrance arises from circumstances outside the character. A mountaineer faces a climb up a sheer wall. A hiker lost in the desert faces heat and vast distances. Both the mountain and the desert present circumstances of the environment.

A third type of conflict or hindrance presents itself when the intention of one character is at odds with the intention of another character. A young boy wants to join the army, but his mother opposes his desire. The exercise in this chapter concentrates on this third type of conflict—conflict between opposing parties—because that is generally the most important and frequently used of the three.

Because such conflict deals with at least two people, it tends to reveal more about human nature than does internal or circumstantial conflict. In fact, many instances of internal and circumstantial conflict are elaborated in terms of conflicts with other characters. Hamlet's indecision—an internal conflict—is conveyed through a series of conflicts with Ophelia, Polonius, Gertrude, and other characters. Willy Loman's internal conflict is dramatized through external conflicts with his wife and his sons.

The *intention* of a character meeting some form of *resistance* produces conflict. To those ingredients must be added another—*outcome*. There must be some resolution to the conflict. In most plays, taken as a whole, the main character either succeeds or fails. A character overcomes the opposing force or is overcome by it.

There are, however, resolutions other than complete victory or total defeat. In *No Exit* we see that the central characters will never achieve their goals, but are destined to pursue them forever in a hellish eternal conflict. Likewise in *Waiting for Godot* we understand that the dream of the characters—the arrival of Godot—will never be realized even though the characters will continue their determined waiting. In both cases the situation is resolved or ended through the recognition by the audience that the goal will prove eternally elusive.

The scenes of a play have the same characteristics of conflict that whole plays have. Characters pursue different intentions, they come into conflict, and the conflict reaches an outcome. However, a scene does not show the entire story. A scene is only a unit of action that dramatizes one segment of the whole story. Similarly, the objective of a character in a scene is usually just a part of the overall intention of the character. A scene represents one incident in a larger whole, and, although it includes a part of the story, it also leaves important elements unresolved. The outcome of a scene—any scene except the last of a play—is not the final resolution or outcome of the conflict.

Within a scene, the outcome proceeds in one of three ways: (1) one character wins and the other loses; (2) the characters compromise; or (3) the characters reinforce their disagreement. But the scene may be only one segment of a much larger conflict. Hence, Hamlet in one scene persuades the players to perform a certain play. He achieves his objective. But his victory is merely one part of a larger intention, and in later scenes he pursues other objectives within that larger intention.

In the recent movie *When Harry Met Sally*, Harry and Sally have an argument. The incident is resolved when they agree to go their separate ways. The scene reinforces their conflict. Neither character, however, is pleased with that outcome, and so the conflict is resumed and concluded in later scenes through a process of compromise and accommodation. It is important to remember that the action of a scene can be resolved or ended even though the basic conflict remains.

I have chosen the word "outcome" to indicate that a conflict of intentions may be ended or continued. Other writers have used other terms, and two of those terms—crisis and climax—have been used so frequently that they require comment. Unfortunately, every author who has used those terms seems obliged to define them somewhat differently from anyone else. Everyday language uses crisis for virtually any kind of conflict. Hence a person trying to make a decision (internal conflict), a person trying to climb over a wall (external circumstance), or a person battling another person (conflict of intentions) all present a scene or moment of crisis. Climax has generally been used to designate that point of the conflict when one side wins or loses decisively or when another form of resolution (such as a compromise or reconciliation) is reached.

The exercise for this chapter asks you to put two characters in a direct conflict. You will notice almost immediately that virtually any two characters can be placed in such a conflict situation. Test this idea yourself. Think of any two people or characters. Then devise some situation where they can come into conflict. There's *always* a way.

To arrange people in conflict, think first of any two people. Next, think of a place or an environment where the two of them might come in contact with each other. For instance, if one of your people were a taxi driver, the location might be a taxi. If one were a lawyer, the location might be a courtroom or a law office. Alternatively, you might select a public place where *any* two people could come in contact with each other: an airport lounge, a baseball game, a lunch counter, or a street corner.

Once you have the two characters in the same place, you need to generate a conflict. That might have something to do with the occupation of one of the characters, such as the route the cab driver takes or the strategy the lawyer devises. In many cases conflicts involve objects. Material items have a way of focusing disagreements. Perhaps the passenger in the taxi wants to eat some particularly offensive or messy food in the car, or wants to transport an item that the driver doesn't want in the vehicle.

In this exercise you will be working with stage terminology and stage directions for the first time. Virtually all plays describe a setting for the action. This is the environment where the scene takes place. George Bernard Shaw liked to write lengthy, entertaining descriptions that were practically comic essays. Earlier in this century Shaw's style was much admired, but few writers possessed Shaw's intellect or his comic skills, and such discourses are now rare.

Shakespeare, of course, almost never described a setting except in the speeches of his characters. Some contemporary writers provide little more than Shakespeare. David Mamet's entire description of the set for *American Buffalo,* which was an elaborate, realistic secondhand store with hundreds of items, read simply: "Don's Resale Shop. A junk shop."

Your own description of a setting should probably be somewhere between Shaw's expansiveness and Mamet's terseness. You need not describe the setting in great detail, but you should mention items that are needed for the action of the piece. You should also include selected elements that establish the proper mood or that succinctly define the characters who inhabit the environment.

Stage directions are normally written for a proscenium stage configuration. A proscenium stage is one of the three kinds of formal stages (see illustration). In an arena, or theater-in-the-round, the audience surrounds the action on all four sides. A thrust stage juts out into the audience. The audience is arranged on three sides of the stage, and the fourth side usually includes a scenic background. In the proscenium arrangement the audience faces the stage from one direction and a curtain or "proscenium arch" hides all or part of the stage and the stage mechanism from the audience.

Theatrical convention, which means normal theatrical usage, calls for stage descriptions to be written for a proscenium stage. In a proscenium configuration the part of the stage closest to the audience is called "downstage," while the part furthest from the audience is "upstage." Those terms are holdovers from a time when stages were typically sloped or "raked" so the audience could see better. In most theatres now the audience area is sloped and the stage is flat, but the conventional terminology remains.

The side-to-side stage nomenclature is defined in terms of the *per-*

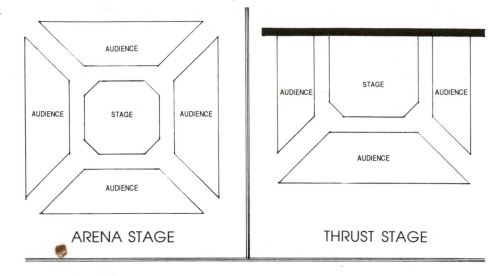

ARENA STAGE

THRUST STAGE

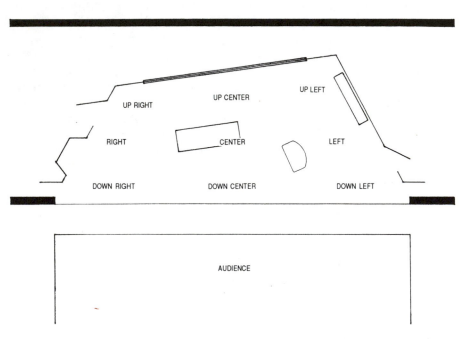

PROSCENIUM STAGE

formers rather than the viewers. Therefore, "stage right" is actually the viewer's left and vice versa. Your description of the setting should reflect those stage conventions.

The stage setting illustrated shows a room with a fireplace on the right wall. A door is up right. There are windows on the upstage wall, a bookcase up left and a second door downstage left. The drawing shows a couch at center stage with a chair to the left of the couch. Remember, you need to specify only what is important. It may be enough to say simply "The living room of a modern house."

As you work on the exercise in this chapter, you will find that characters begin to emerge in greater detail as they react within a specific environment, speaking to and struggling with other characters. That's a natural development from a conflict, and the next chapter will deal more specifically with characters and with various ways in which characters evolve.

In this same exercise you will also be writing dialogue for the first time. A later chapter will go into more detail about dialogue, but for now it is more important for you to visualize and activate a conflict. Let what the characters say derive from what they want and from the kind of people they are.

At some point you may consider writing a screenplay instead of a play. There are many similarities in the two forms. Both plays and screenplays dramatize action. They both use conflict, character, and dialogue in similar ways.

There are, however, important differences. Although plays can certainly jump from place to place, they usually concentrate on one place and one set of characters for longer periods of time than do screenplays. I believe that a great part of the process of dramatic writing is learning to work with characters within particular situations, and I think you learn more about such fundamentals as character, dialogue, action, and conflict by working within the confines of a stage format because it forces you to explore relationships in greater depth. Lillian Hellman, who wrote in many different formats, said that plays focused on one event more than other forms of writing, and I think it is that attention to one event that helps the writer probe.

Screenplays also rely far more than plays on the juxtaposition of visual images, particularly close-up visual images. A camera can show you a dripping faucet, followed by a gun lying on a floor, followed by drops of blood, and finally a corpse. Then it can cut to a man picking his teeth in a car. That kind of flowing imagery makes screenplays more like short stories or novels than plays.

Plays and screenplays employ very different writing formats, but that problem is easily overcome by looking at a book about screenwriting. If at some point you wish to tackle screenplays, I've listed several instructive books in the bibliography, but for the exercises in this book I would encourage you to stick with a play format.

Now let's proceed to the second exercise.

● ● ● **EXERCISE 2**

Place two characters in a scene of direct conflict. Write the scene in stage terms, indicating how the scene should be staged. Use dialogue as needed. Bring the conflict to an outcome and end the scene. Write a minimum of two pages.

The following scene is an example of a two-person conflict scene written by a student in a playwriting class. Except for a few instances where I have used a condensed format for a few lines, all examples in this book use the accepted playscript format of the following scenes. The NAME of the person speaking is capitalized and centered on the page. Stage directions are indented almost to the center of the page, and in the stage directions CHARACTERS' NAMES, ENTRANCES and EXITS are customarily capitalized. You should get into the habit of using that format.

EXAMPLE

SIS

by Anne Dodson

> (JOANIE is sitting on her bed in her room
> studying when her sister KATHY WALKS IN.
> JOANIE looks up.)

KATHY

Joanie, can I wear your green sweater? It'll go just
right with my yellow turtleneck.

> (She begins to go through the drawers of
> JOANIE'S chest.)

I'm going out with Errol tonight, and I really want to
look good.

> (She finds the sweater and starts to leave.)

Thanks.

JOANIE

Wait a minute, Kathy. I never said you could wear my
sweater and here you are, taking it anyway.

KATHY

> (At the door, she checks sweater for pulls
> and lint.)

Yeah, I know. Look, you've worn plenty of my clothes
without even asking me. At least I ask first.

JOANIE

You call that asking? You come in here, you practically
tell me you're going to wear my most expensive sweater,
and you call that asking? Well I'm sick of it Kathy.
You're always doing that to me and if you don't cut it
out I'm gonna. . . .

                    KATHY
Oh shut up and leave me alone, Joanie. You sure are in
a bad mood tonight. I'm going to be late, so just shut
up. I'm wearing this sweater whether you like it or
not.
          (JOANIE starts to speak.)
Besides, you owe me from last weekend when you wore my
red blouse and ripped it and didn't even tell me. But I
guess you don't remember that.

          (KATHY EXITS. JOANIE remains seated on the
          bed, then yells after her sister.)

                    JOANIE
Okay, Kathy, go ahead and wear my sweater, but you're
gonna wash it as soon as you get home. Do you hear me?

                    KATHY (Off)
Oh, shut up!

          (JOANIE falls back on her pillow.)

                    THE END

This is a somewhat more subtle scene from another student's.

EXAMPLE

                    RECKONING
                 by Julie Williford

THE SETTING is the living room of a small
house. The furniture is old but comfortable.
Small antique pieces are mixed in with more
modern furniture. The floor is hardwood.
There is a fireplace with a fan-like cover.
An upright piano is situated against one
wall. A couch, the cover slightly faded,
lines another wall. A wooden coffee table is
in front of the couch. Antique colored
glasses sit on a silver tray on an end table
next to the couch. At the other end a glass-
based lamp stands on a similar table. The
room is dark, even though light is coming
through the window.

Two women ENTER from a hallway stage left.
One is old, about 70, and she's wearing a
print polyester shirt outside of stretch
pants that are too short for her. The other
woman is about 40. She enters behind the
older woman. She's wearing a light sweater

and slacks, and she's carrying a small note
pad and pen.

MOTHER

Now for this room. . . .

BETTY

Mother, I really don't see the point to all this.

MOTHER

Betty, if it's too much trouble for you to do this
little thing for me, I'll do it myself.

BETTY

It's no trouble, but I can't understand why you find
this necessary.

MOTHER

It will be a comfort to me to know that my possessions
will be looked after when I'm gone.

BETTY

Mother, you have plenty of time for this. The doctor
told you you're in perfect health.

MOTHER
            (Puts her hand to her chest)
Doctors. What do they know.
            (Coughs slightly)
Don't you think I know more about my body than any
doctor?

BETTY

Dr. Phelps has been seeing you for 25 years, Mother. He
knows your body pretty well.

MOTHER

Are you going to help me or not?

BETTY

All right, Mother.
            (Lifts pen and pad to write)

MOTHER

Now in here . . . you did write down that the butler's
desk from the guest room is for Michael?

BETTY

Yes, Mother, it's right here. (Reads) "Butler's desk
for Michael."

MOTHER

Good. Now in here . . . the piano is to go to the First
Presbyterian Church on Barker Street.

                    BETTY
          (Writing)
I know where it is, Mother.

                    MOTHER
Good. And the end tables . . .
               (She goes over and strokes one of the
               tables.)
. . . it's beautiful wood, isn't it? Well, they go to
your Aunt Mary. She always loved them.

                    BETTY
Aunt Mary. Yes, Mother.

                    MOTHER
And the water glasses go to Cousin Alma.

                    BETTY
But Mother . . . I always wanted those glasses. I told
you that. Why do you want to give those pretty glasses
to stuffy Cousin Alma? You don't even like her!

                    MOTHER
Please, Betty. Cousin Alma was very good to you when
you were little. You shouldn't speak about her that
way.

                    BETTY
Cousin Alma is a stingy old snob and you know it! The
glasses are for me.
               (She writes it down.)

                    MOTHER
If you're going to make such a fuss over them . . .
alright. I'll just have to find something else for
Alma. Those glasses would be perfect, though.

                    BETTY
          (Firmly)
We'll find something else for her.

                    MOTHER
I said all right. Now I think that's all from this room.
               (MOTHER HEADS OUT the same hallway she
               entered.)
You did write down that the butler's desk is for
Michael, didn't you?

                    BETTY
          (FOLLOWING HER OUT)
Yes, Mother. "Butler's desk for Michael."

                    THE END

EVALUATION

There are several instructive questions you might ask about these two scenes. First, do we know what the conflict is? Can we tell what each character's objective is? In other words, are the objectives clear and in opposition to each other?

Second, what do we discover about the characters through this conflict? Do we care about the characters? Do you want one character or the other to win?

Third, is there action in the scene? Can you state in a phrase what we see the characters doing?

Fourth, is the setting significant to the scene?

And finally, does the dialogue sound like people actually talking? Is the language appropriate for the characters in the scene? Do the characters talk differently?

In the first scene there is certainly no earth-shattering meaning, but the conflict is direct and the situation is one with which many people can identify. Kathy wants her sister's sweater, and Joanie doesn't want her to have it. The objectives are clear and in opposition. To her credit, the author bases the scene on a visual action—getting and taking the sweater—and the verbal argument stems from and complements that primary action.

The setting is simply stated as Joanie's bedroom. Although there are no elaborate details, it is significant that Kathy invades Joanie's space. Also, as the action develops, a chest or closet for the sweater is clearly indicated.

The author has provided several elements that make this brief conflict interesting. At first we, as the audience, join with Joanie in taking offense at Kathy's unwarranted intrusion into her space—Kathy doesn't even knock—and at Kathy's cavalier attitude toward property rights. But the author allows Kathy some telling responses. Her sister also borrows clothes and—the deciding blow—has ripped Kathy's blouse recently. We may still be uneasy about Kathy's domineering attitude, but we are satisfied that she has at least a decent claim to wear Joanie's sweater.

Although the characters use a similar vocabulary, they are certainly differentiated. Kathy's sentences are shorter and to the point ("Oh, shut up"), while Joanie uses many words to little effect. Even though the author does not say so, she creates the impression that Kathy is the older sister. She is definitely more authoritative. Joanie, in comparison, seems immature. You can practically hear the whine in her complaints.

In fact, Joanie's complaints and Kathy's deft responses place the conflict in a larger perspective. This small incident suggests other significant possibilities. Perhaps Joanie is jealous of Kathy's date or envious of Kathy's maturity. In any case the retorts allow the audience to view the scene in terms of eternal sibling rivalry. Kathy wins this round of that never-ending conflict, and the final lines show Joanie trying to salvage a little of her pride ("Go ahead and wear my sweater, but you're going to wash it . . . ") and Kathy punctuating her victory. The scene is resolved although we suspect the conflict will go on and on.

The second scene is more complex. The mother's objective is quite clear. She wants to document who is to get her possessions after she dies. The position of the daughter, however, is less distinct. She implies that she doesn't want to make the list, but she does it anyway. She tries to convince her mother that she's still healthy, but the mother nicely dismisses her argument with the non sequitur "Are you going to help me or not?" The mother says, in effect, "I don't want to talk about that," and she places her daughter in a guilt situation: "If you don't help me, you'll feel guilty." The mother also uses guilt as a tactic earlier when she says "If it's too much trouble for you to do this little thing for me, I'll do it myself." In both cases the daughter gives in. But Betty's quickness to speak up for what she wants in some measure belies her unwillingness to make the list. Perhaps she's just as glad as her mother to get on with it and her protestations are merely polite. Perhaps the mother really wants her daughter to speak up for what she wants but realizes Betty would never do that directly. Perhaps the real conflict is Betty's inability or unwillingness to accept the inevitability that her mother is aging and eventually will die. There are various emotional levels here, and the author approaches them obliquely without the direct argument that engages the two sisters.

The dialogue moves gently between mother and daughter. However, some of it, such as the lines "I can't understand why you find this necessary" and "It will be a great comfort to me," are unnecessarily formal.

Whereas the set for "Sis" could be virtually any bedroom, this room calls not only for certain specific items but also for a particular shadowy mood, a somber, closed-off, unlived-in atmosphere.

As they created conflict, both writers also began to generate characters. Now that you've seen how any two characters can come into direct conflict, let's take a more detailed look at exactly how characters develop.

3

• • • • • • • •

Character

EXERCISE 3 • CHARACTER
List the characteristics of three people you know very well. Select one of those people and place a character based on that person in a situation of conflict with another person. Use correct stage terminology and dialogue as needed.

In the previous chapter I stated that conflict does not arise out of nothing; it stems rather from intentions at cross-purposes. Intention, in turn, stems from character. A logical question follows: If conflict comes from intention and intention comes from character, why start with conflict? Why not start with character? Many writers, as you can see from the Character vs. Action chart, ask just that question and argue for the preeminence of character. Lajos Egri, a significant and sensible writer on playwriting, concluded, "There is no doubt that conflict grows out of character." Why not, then, begin there? The answer is this: Character does not necessarily produce conflict, but conflict necessarily brings out character.

Let's assume you have thought of two characters for the conflict exercise—a doctor and trash collector, for instance. What do we know about them? Nothing. Now let's put them in a situation of conflict. Using the prompts suggested, we can first envision a place where the two characters come together, and we will use the occupation of the trash collector. He's making his rounds, and he comes to the house of the doctor. Now imagine an object that could force their conflict.

The trash collector is picking up bags, and a plastic bag bursts, littering debris over the driveway in front of the doctor's expensive house. The trash collector sighs and mutters an expletive under his breath. Suddenly a man emerges from the house and hurries to the trash collector.

Doctor: Hey, what are you doing here?

Trash collector: The bag broke.

Doctor: I can see that. Are you just going to leave
    this garbage all over my front yard?

CHARACTER VS. ACTION

For Character	*For Action*

For Character

The dramatist who hangs his characters to his plot, instead of his plot to his characters, ought himself to be hanged.

John Galsworthy (paraphrased)

My plays deal with people, and thinking, and believing and philosophizing are all, to some extent at least, a part of human behavior.

Friedrich Duerrenmatt

Before I write down one word, I have to have the character in mind through and through. I must penetrate into the last wrinkle of his soul. I always proceed from the individual; the stage setting, the dramatic ensemble, all of that comes naturally and does not cause me any worry, as soon as I am certain of the individual in every aspect of his humanity.

Henrik Ibsen

The difference between a live play and a dead one is that in the former the characters control the plot while in the latter the plot controls the characters.

William Archer

Dream out a story about the sort of persons you know the most about and tell it as simply as you can.

Attributed to Owen Davis

There must be a force which will unify all parts, a force out of which they will grow as naturally as limbs grow from the body, [and that force is] human character, in all its infinite ramifications.

Lajos Egri

Every great literary work grew from character . . . character creates plot, not vice versa.

Lajos Egri

For me . . . ideas start with character.

John Van Druten

For Action

Everything hangs on the story; it is the heart of the theatrical performance. For it is what happens *between* people that provides them with the material to discuss, criticize, alter.

Bertolt Brecht

The dramatist must be by instinct a storyteller.

Thornton Wilder

Things occur to me first as scenes with action and dialogue, as moments developing out of their own vitality.

George Bernard Shaw

Plays should deal with moments of crisis.

Marsha Norman

A play lives by suspense, and suspense comes from complication.

Kenneth MacGowan

History shows indisputably that the drama in its beginnings, no matter where we look, depended most on action.

George Pierce Baker

Wherever you start, eventually the material must take on some sort of shape. In order to give it shape, you have to get some type of story.

Josephine Nigli

Trash collector: Look, Mac, you got a million dollar house there, so how come you use cheap bags? It ain't my fault it broke.

Doctor: I got it out here, didn't I? It didn't break on me.

Trash collector: I don't need this, buddy.

Doctor: You get it cleaned up. Every bit of it. I've got to take out a gall bladder in twenty minutes and I can't even get my car out of the driveway.

Trash collector: So call a cab.

Doctor: You get it picked up or I'll report you!

Trash collector:
> (Throwing down the remnants of the bag in his hand and signaling for the truck to pull on.)
Pick it up yourself, Mac. And next time use a better bag!

In the course of this small conflict two characters—an arrogant doctor and a proud, defensive trash collector—begin to emerge.

As this example illustrates, you cannot create a dramatic character in isolation. You may list a multitude of characteristics of your characters: age, occupation, physical appearance, favorite activity, recently read books, likes and dislikes, and so on. But the characters won't really begin to reveal themselves until you place them in dramatic situations rife with action and conflict.

Placing a character in a particular situation is actually one of the oldest techniques of playwriting, for through that process the character is revealed. The playwright begins to understand more and more about the characters being created, and, in the end, the audience will come to understand those characters as well.

The revelation of character through conflict and action in drama should not come as a surprise, for the process is quite similar to what occurs in real life. You don't simply meet a person and instantaneously know that person's character. Rather, the character of individuals emerges bit by bit through their actions, through what they do and say, and through their interactions with other people. In just the same way character emerges in drama. Character, after all, is one of those things that is difficult to approach directly but is vividly revealed through what happens.

Let us for a moment consider *you* as a character. Imagine that you're driving down a winding two-lane highway. A bystander would observe aspects of your personality. Are you driving a sparkling new Volvo or a rusted pick-up truck? Are there any bumper or window stickers espousing causes, identifying a school, or indicating a special parking permit? Are you wearing jeans or shorts or more dressy attire? Do you drive cautiously on the turns, or do you attack them like a racecar driver? All of those elements provide clues about you.

Now let's complicate your life. Suppose that the road is icy. Do you sit up in the driver's seat? Do we read concern on your face or are you calm? Perhaps you relish the challenge. What if another car cuts you off? Do you curse or make a gesture at the other driver? Or do you just shake your head, take a breath, and forget it? Imagine that your car breaks down or has a flat tire. Do you know what do do to fix the problem yourself? Do you hail a passing motorist? In these instances the conflict generated by the weather, another motorist, or the inanimate automobile itself, and your responses to those challenges, would reveal to your audience additional facets of your personality.

As you saw from the Character vs. Action chart, critics have expended an enormous quantity of ink in a long-running debate over whether character or action is more important to a play. No one has expressed the case for the dominance of characters over action more eloquently than John Galsworthy, the author of early twentieth-century plays including *Justice* and *Strife*. He wrote:

> In drama, undoubtedly the strongest immediate appeal to the general public is action. . . . The permanent value of a play, however, rests on its characterization. Characterization focuses attention. It is the chief means of creating in an audience sympathy for the subject or the people of the play.

More succinctly, Galsworthy said, "A human being is the best plot there is." That point of view has many supporters, including Henrik Ibsen, Lajos Egri, Friedrich Duerrenmatt, William Archer, and John Van Druten.

But if writing plays were a debate among authorities, the case for the preeminence of action or plot could be argued by Kenneth MacGowan, Bertolt Brecht, Thornton Wilder, Josephine Nigli, and others. Wilder, for instance, concluded that "drama on the stage is inseparable from forward movement, from action."

Unfortunately, as is the case with other long-standing questions such as "Which came first, the chicken or the egg?" the question "Which is more important, character or action?" is, finally, an empty one. Character and action are inseparable. It is as impossible as it is undesirable to have one without the other.

A scene or a play can begin with a story or it can begin with an interesting character. Moreover, a play can begin in many other ways as well. It can start with a word or a phrase. It can be propelled by a firmly held conviction. Tina Howe once said that she always starts a play with a setting. Her play *Painting Churches* takes place in a Boston interior while her *Coastal Disturbances* takes place on a beach.

Wendy Wasserstein has said that her plays often spring from a visual image. Her Pulitzer Prize–winning piece *The Heidi Chronicles* began with the picture of a woman standing in front of a group of other women and saying that she's never been so unhappy in her life.

What's important is not where plays begin, but where they end. No matter what they start with, somewhere along the line of their development they must incorporate all the elements of an effective play—well-defined characters using interesting dialogue in a forward-moving story, all brought together within an enticing environment.

To demonstrate how different elements influence each other, particularly how action and character reinforce one another, let me describe an action. This action was written by a student in one of my playwriting classes.

EXAMPLE

```
         A WOMAN AT CHRISTMAS
           by Mary Parker

A middle-aged woman stands scrutinizing a
Christmas tree, which stands undecorated in
the corner of a small room. Deciding that it
is leaning toward the left a fraction, she
walks determinedly to it, kneels and presses
the trunk toward the right while she
tightens the screws of the stand on the left
side. She stands, takes a few steps
backward, runs her fingers through her
graying hair and observes the tree for a few
seconds. Sighing, she walks slowly from the
room and returns with a stack of boxes. As
she stoops to place the boxes down in a
chair beside the tree, the top box falls off
and shattering glass is heard as it hits the
floor. She sets the other boxes down and
then bends slowly over the fallen box. She
breathes in deeply and removes the top of
the box. Upon seeing the contents, she sits
on the floor and exhales.

Hesitantly she reaches into the box and
brings forth a large broken piece of shiny
green glass. She holds it up toward the
light and watches the light sparkle off the
broken edges. She places it on the floor by
the box and gets up. She walks out of the
room without looking back and slowly shuts
the door behind her.

              THE END
```

That action not only exudes a somber tone, it con character. The middle-aged woman is alone at the holiday ing expression and regular pace alert us to her depression that the tree be properly upright informs us that this is a lous woman. The scene suggests that she is attempting to keep a hold on her lonely life through the ritual of trimming the tree. But when the ornaments fall and a special ornament is broken, she becomes desolate and cannot sustain the pretense. She gives up and leaves the room.

Now let's turn around and describe a character. I'm thinking of a boy of about ten. He has a constant runny nose, which he wipes with the back of his hand. This young man wants to be thought of as tough. Even on the coldest days he leaves his shabby coat unzipped. He is something of a bully, taunting the younger children at school and throwing snowballs at them.

Here we have the seed of a character, and even in this early stage the character is beginning to emerge in terms of action and conflict: What he wears and how he wears it; his personal behaviors; what he does to other people.

You probably discovered as you wrote your action and direct conflict scenes, that they, like the paragraph about the woman at the Christmas tree, began to suggest moods and convey character. That occurs not because action and conflict are more important than other elements, but because the various elements of drama work together, support each other, and reveal each other.

That is what happens in the action of the women trimming the tree. Her determined walk and unsmiling visage work together with her gray hair to suggest a certain character. That in turn is supported by her slow, hesitant movement and careful breathing in response to the fallen ornament box. And all of those elements, juxtaposed within what should be a joyous event, create a revealing, dramatic moment.

Although you cannot create character in isolation from other elements, you can use characters as the impetus for a scene or a play. One way to do that is to write about characters with whom you are familiar.

• • • **EXERCISE 3**

Think of three people who are close to you. For each one, make a list of physical characteristics. List any particular vocal characteristics or particular modes of speaking. Add to your list other external information such as age, occupation, religious affiliation, race, and origin. List particular likes and dislikes of the person. Then add psychological factors that reflect the person's values. What bothers this person? What does he or she care the most passionately about?

When you have completed three lists, select *one* of these people and place a character based on this person in a situation of conflict with another person.

EXAMPLE

RISK

by Robin Jackson

THE SETTING is a small living room. It's
winter, and a fire burns in the fireplace.
There is a coffee pot and cups on a small
side table.

At center is a card table at which two
people are absorbed in playing Risk. They
are a young couple in their early twenties.
DAVID, the man, is winning. He owns the
whole board except for two countries. GAIL,
the woman, sits opposite him, rolling the
dice unconcernedly.

DAVID

Yakutsk to Mongolia.
            (They throw dice.)
Six, six.

GAIL

Five, five. Your country.

DAVID

Great. Now, where shall I attack next?

GAIL

Why not Egypt? I mean, I know you already own it, but
you can afford the armies.

DAVID

No, I think I'll take India. I'm into gauze.

GAIL

I don't know. A little civil war would be kind of
exciting.

DAVID

You might hold me off. You still have two armies, so
you can roll two dice.

GAIL

I prefer to play my own game. I'll roll one, thank you.

DAVID

OK, but you can't prolong it much. I still have five
cards to turn in. All right, Middle East to India.
            (They roll.)
Five, four.

                    GAIL
Two. I lose one army, and I retire undefeated.

                    DAVID
You what?

                    GAIL
I quit. Undefeated.

                    DAVID
Roll the dice.

                    GAIL
No.

                    DAVID
I'll make you roll.
               (DAVID grabs for her wrist. GAIL calmly tips
               the table. The Risk game and its armies
               slide into DAVID's lap.)
That was real childish.

                    GAIL
               (Moves to the coffee pot)
So sue me. I'm going to get a cup of Sanka and go to
bed. You can pick up the game.

                    DAVID
No you're not. I demand a rematch.

                    GAIL
Not tonight. It's almost two.

                    DAVID
So? Either admit you lost or play me another game.

                    GAIL
I retire, undefeated, and I'm going to bed.
               (GAIL starts out door.)

                    DAVID
Not in my house!

                    GAIL (Angrily)
Your house? Since when did it become your house. I make
half the payments, you know.

                    DAVID
I'll give you a refund tomorrow.

                    GAIL
The hell you will. I think that's pretty low, since you
started the whole business.

                              DAVID
I started it? You quit!

                              GAIL
You broke the nonaggression pact between North Africa
and Brazil. We were supposed to fight in Europe. That
was the pact. But no-o-o, you come stomping through
North Africa all the way down through Madagascar. So
don't go telling me I started it. I just got my
revenge.

                              DAVID
What a bitch!

                              GAIL
At least I was loyal to the pact. You can't even say
that.

                              DAVID
I can say I'm a better loser than you are.
                (GAIL throws her cup at DAVID.)
All right, that's it. Get out of my house!

                              GAIL
It's my house, too!

                              DAVID
Not anymore! I'll give you your payments back in the
morning. But now, get out! Don't bother to pack. Just
get the hell out the door!

                              GAIL
David, it's two in the morning. It's seventeen degrees.

                              DAVID
I'll send your things somewhere tomorrow.

                              GAIL
There's three feet of snow out there. I can't walk in
three feet of snow!

                              DAVID
Then tunnel!

                              GAIL
David!
                    (GAIL STORMS FROM THE ROOM, slamming the
                    door behind her. We hear another door slam.
                    DAVID stares after the door, then shakes
                    himself and runs after her.)

                         DAVID
Gail! Wait, Gail.
                (DAVID EXITS and we hear the second door
                slam.)

                       THE END

     The preceding scene used a boyfriend for a spark. The author of
the following scene found her characters in a mother and a daughter.

EXAMPLE
                     THE SUIT
                  by Daryl Harrison

     THE SETTING is the dressing room area of a
     department store with several curtained,
     boring-beige stalls. Brightly colored pieces
     of clothing are lying on the floor, draped
     over the stalls, and hanging on the reject
     rack, right.

     A WOMAN, fortyish and dowdy, waits outside
     one of the stalls. She has several bathing
     suits on hangers draped over her arm.

                        MOTHER
Do you have something to show me?

                       DAUGHTER
                (From inside a curtained stall)
I've almost got it on.

                        MOTHER
Well, hurry up. I don't want to fight that traffic.
                (A 16-YEAR-OLD GIRL emerges from the stall.
                She is wearing a black print bikini.)

                       DAUGHTER
                (Brightly)
Well? What do you think?

                        MOTHER
Too skimpy.

                       DAUGHTER
Mom. . . .

                        MOTHER
Besides, you're too young to be wearing black. Let's
see another one.

                    DAUGHTER
That's the last one.

                    MOTHER
It can't be. What about that navy one. You haven't
tried that one on yet.

                    DAUGHTER
I don't like it, Mom. It has a skirt.

                    MOTHER
What's wrong with a skirt? It's reminiscent of the
1950s. You'll look like a bathing beauty of the 50s.

                    DAUGHTER
I'll look like the nerd of the 90s. Nobody wears
skirts. What else did you find?
          (She looks through the suits on her Mother's
          arm.)

                    MOTHER
This one's really cute.

                    DAUGHTER
It's terry cloth. It expands when it gets wet.

                    MOTHER
How about this one?

                    DAUGHTER
Mom, that's hideous!

                    MOTHER
I forgot you don't like the padded cups. What do you
think of this?

                    DAUGHTER
I kinda like that. I don't remember seeing that on the
rack.
          (She takes it and ENTERS THE STALL and
          closes the curtain.)

                    MOTHER
It was mixed in with the elevens.

                    DAUGHTER
Oh.

                    MOTHER
That's going to have to be the last one. I want to be
home before rush hour.

                    DAUGHTER
I'm hurrying.
              (EMERGING FROM STALL)
Can you help me with the straps?

                    MOTHER
              (Adjusting strap)
There. Now let me see you.

                    DAUGHTER
I like this one.

                    MOTHER
It looks very nice, dear. And it's lined, too.

                    DAUGHTER
Can I get it?

                    MOTHER
Yes. Now get your clothes on.
              (DAUGHTER RETURNS TO THE STALL and closes
              the curtain.)
Tell me how much that is, and I'll write out the check.

                    DAUGHTER
It's forty-eight dollars.

                    MOTHER
No, dear, the sale price.

                    DAUGHTER
That is the sale price. It's down from sixty.

                    MOTHER
Well, I'm sorry, but I'm not paying forty-eight dollars
for that tiny piece of fabric. Are your clothes on? We
need to start home.

                    DAUGHTER
              (COMING OUT of the stall)
But Mom, I have to find a suit for tomorrow.

                    MOTHER
              (Taking suit from daughter and putting it
              with others on the reject rack)
We'll look when the prices come down. You can wear last
year's.
              (MOTHER LEAVES dressing area.)

                    DAUGHTER
              (FOLLOWING)
This is the end of my social life.

                    THE END

EVALUATION

In the first example the student used a boy she knew as the germ for the scene. The key for her development of the character was the way his overcompetitiveness sometimes blinded him to other things he should have cared about. When I first saw the scene I'd never played Risk, and even though an audience couldn't actually see the armies or the dice and might not know the exact meaning of each reference to the game, the game still provides an effective metaphor for examining the characters and their relationships. The fact that the game is "Risk," with all that implies, even though the name is never mentioned, is a bonus. We sense, for instance, that something more than just this game is troubling this couple.

Writers often use games in plays to provide dramatic tension, to act as a metaphor, or to help provide a structure. The central focus of *The Gin Game,* the Pulitzer Prize–winning play by D. L. Coburn, is an on-again, off-again card game played by an elderly man and woman. In Tina Howe's *Road to Zanzibar*, the characters play a game called "Geography," which leads to the title of the play. In *A Streetcar Named Desire,* Stanley's poker night serves as an emotional backdrop for the climactic moment of the play.

The author of the second example used her mother as the impetus for the scene. Without our being told directly, we discover that the mother is conservative about style, perhaps even prudish, and she's careful about her money. She examines things thoroughly, even finding a swimsuit in the wrong section of the rack. She wants her daughter conservatively dressed, and she doesn't want to spend a lot of money. Because those desires come in conflict with the desires of the daughter to have a swimsuit that is modern and sexy, characters emerge.

The selection of the item over which the conflict occurs—a swimsuit—is a nice touch. There is something particularly revealing about people's attitudes toward garments that uncover parts of the body. If the mother and daughter disagreed about an everyday outfit, that would not be nearly as intriguing.

The weakest part of the scene is the girl's reason for wanting the new suit, which is only hinted at in the final lines. The girl's desire for the suit could be introduced earlier and substantially strengthened. That, in turn, would add to her final disappointment.

On the other hand, this little scene does two other things very nicely. First, it puts time pressure on the situation. The mother wants to get on her way before the rush hour.

Second, it seems to go in one direction, and then it changes. In dramatic terms, that is called a "reversal." It appears that the mother and daughter have actually agreed on a particular suit, but then that suit winds up being too expensive, and they leave without a swimsuit.

In both scenes characters are revealed because of what they want and how they go about getting it. Not only does the boy want to win the game, he also wants the girl to lose. He wants her to admit that he's better. Not only does the mother want her daughter to get a swimsuit of which she approves, but she also wants her to get it for a decent price and within a limited amount of time.

In the next chapter we'll work on how to provide characters with effective voices—in other words, how to write dialogue.

4

• • • • • • • • •

Dialogue

EXERCISE 4 • DIALOGUE
Listen to and write down the dialogue of two selected individuals. Write a scene of conflict in which you incorporate into your dialogue some of the speech habits you've observed.

Just as character arises out of action and conflict, so too, dialogue stems from conflict and action and character. It does not just exist by itself. If you try simply to create characters talking, they will have little to talk about and the resulting dialogue will veer toward pretentious philosophizing or empty small talk. If, on the other hand, you create the dialogue to serve a purpose, to complement an action, or to gain a victory in a conflict, it is more likely that it will seem necessary and important.

In some ways dialogue is a trap. Because we are so accustomed to think of plays as dialogue, we run the danger as writers of concentrating only on the words and forgetting the significance of the action, the conflict, and the characters. Good dialogue should be a means for a character to accomplish an end, so don't lose sight of the goals your characters mean to achieve.

Some playwrights have particular difficulty writing dialogue. Instead of writing as people speak, they write as if they were penning an academic composition: complete sentences, proper grammar, no contractions. Playwriting students often seem to write dialogue more formally than they write term papers.

There are, of course, some basic guidelines for dialogue that can assist beginning writers. Konstantin Stanislavsky, the famous Russian director and the man credited with developing an approach to realistic acting, once said that all he wanted was for performers on stage to act like people in real life. That sounded simple, but somehow it turned out to be more difficult to achieve than it sounded.

Writing dialogue is much the same. All we want for realistic dialogue is that characters in plays talk the way people talk in real life. That means they use contractions, incomplete sentences, repetitions, fragments of thought, perhaps bad grammar, and all the other imprecisions of expression to which we humans are subject. The doctor and the trash collector in their brief encounter in the previous chapter used contractions ("I got it out here, didn't I?"), fragments ("Every bit of it"), and bad grammar ("It ain't my fault").

Larry Shue, the comic playwright of *The Nerd* and *The Foreigner*, observed how people begin to say something one way, then reverse themselves and formulate the words another way around instead. In *The Nerd*, Rick Steadman has just made a mistake in a game when Tansy says to him, "You can't—maybe I didn't explain this; see, you're supposed to say something beginning with an 'E.' " In another place a character asks: "What're we—? What's the point of this?" Those stops and starts mirror the way people actually put thoughts together.

Different characters should talk in different ways, just as the doctor and the trash collector did. "I don't need this, buddy" is a line only the trash collector would say, not just because of the content, but because of the familiar use of "buddy." And "Are you going to leave this trash all over my front yard?" could only belong to the doctor, not just because of the content, but because of the formal construction of the sentence.

Of course, such things as level of education, intelligence, and environment affect expression. But even with two characters from the same area who are roughly equivalent in intelligence and education, there should be differences. One person may be assertive or optimistic ("Let's open a store!") while another is submissive or pessimistic ("Maybe we could open a store?"), and those qualities should be reflected in their speech.

Remember, too, that people in real life do not generally talk in speeches. Just as effective acting demands genuine interaction between the performers, so dialogue should be at least a two-way conversation, with a give and take between the characters. The doctor responds to "It ain't my fault it broke" with "I got it out here, didn't I?" Each statement becomes a stimulus causing a response that in turn becomes a stimulus causing a response and so on.

David Mamet is praised for writing realistic contemporary dialogue. In fact, he has been known to frequent restaurants and record the way people actually talk. The result is halting dialogue that seems to move around a subject, dialogue that allows performers to supply nuances to words through the way in which the lines are delivered. This opening exchange from Mamet's *American Buffalo* illustrates his technique. Don operates Don's Resale Shop where the action is set. Bob is a young friend whom Don has sent to follow a certain man.

```
Don: So?
          (Pause)
     So what, Bob?
          (Pause)

Bob: I'm sorry, Donny.
          (Pause)

Don: All right.

Bob: I'm sorry, Donny.
          (Pause)
```

Don: Yeah.

Bob: Maybe he's still in there.

Don: If you think that, Bob, how come you're here?

Bob: I came in.
> (Pause)

Don: You don't come in, Bob. You don't come in until
> you do a thing.

Bob: He didn't come out.

Don: What do I care, Bob, if he came out or not? You're
> s'posed to watch the guy, you watch him. Am I
> wrong?

Bob: I just went to the back.

Don: Why?
> (Pause)
> Why did you do that?

Bob: 'Cause he wasn't coming out the front.

Don: Well, Bob, I'm sorry, but this isn't good enough.
> If you want to do business . . . if we got a
> business deal, it isn't good enough. I want you
> to remember this.

Bob: I do.

Don: Yeah, <u>now</u> . . . but later, what?*

Mamet's ear for the repetitions, juxtapositions, and non sequiturs of human communication provide a natural, flowing sound.

Some individuals will undoubtedly display a better sense of words and the way people use them than others. That's to be expected, just as some people will demonstrate a finer facility for devising plots than others. Eugene O'Neill in the early stages of his career gained fame for writing realistic dialogue that captured the accents and the flavor of the seamen who populated his first plays. Later in his career he was criticized for overwriting his dialogue and for lacking a poetic sense of language. Yet some of his later plays, such as *A Long Day's Journey Into Night,* are among his most powerful. Tennessee Williams' poetic use of words and rhythms within the context of action and character is one of his greatest assets. In any case, don't think you're automatically a playwright because you discover you have a facility for dialogue, and, conversely, don't despair of writing plays because your dialogue doesn't sound quite right.

One of the hardest tasks about writing dialogue is gauging how much information to include, for almost every play needs to inform the

*By David Mamet. Reprinted by special permission.

audience about certain facts. Background information about characters or plot is called "exposition," and good playwrights try to include exposition as naturally as possible. Also, as the play progresses, the author must reveal various elements of the plot or story to the audience. There, too, the aim is to provide needed information as part of the natural flow.

Playwrights often get caught trying to cram too much exposition into a character's line, and the result can be wooden, information-laden speech. I call that problem "chunking" because information comes out in large chunks rather than in bits and pieces.

Look at this opening passage to a scene in a girl's bedroom. There are two twin beds in a carefully arranged room. Melanie, a small woman about 40 who is dressed in black, walks slowly around the room picking up and looking at several items. Finally, she picks up a stuffed bear from one of the beds just as Sandra, also a small woman in black, enters.

Sandra: Hey, what's wrong? How come you came upstairs?

Melanie: It's just so crowded down there, and most of those people didn't even know Daddy. They just came to watch us and see if we'd cry.

Sandra: You sound bitter.

Melanie: I shouldn't have come home. I always get crazy when I come back.

Sandra: Maybe if you'd stayed around long enough you'd have seen it wasn't so bad. You come home <u>knowing</u> everything would be bad, so it would be. Once you got out you never came back just to <u>observe</u> the situation. And off you'd go again.

Melanie: Sandy, that's just not fair. You know it's hard for me to be here. I remember all the times Mom and Dad fought. Sometimes I can still hear them screaming. It never got any better, Sandy. I hate coming here. There have been times, though, when I've wondered if I should have stayed so you could have left.

Almost every line is packed with information. We really don't need to get everything all at once. In addition to the chunking, the speeches are quite long with no give and take, and the emotions are very directly confronted. Those problems give the scene a feeling of wooden characters spouting mechanical lines. Let's look at the same scene with the emotions approached somewhat less directly and the lines broken up and varied.

Sandra: I figured I'd find you here.

Melanie: I've always liked this room.

Sandra: You found Jelly Bean.

Melanie: I don't believe they kept him all these years.
      (Pause)

Sandra: Crowded downstairs, huh?

Melanie: Most of them didn't even know Daddy.

Sandra: They're just trying to be nice. They want to make sure we're ok.

Melanie: They want to watch us grieve.

Sandra: Mellie.

Melanie: It's true. They just want to see tears.

Sandra: Well, they wouldn't see any from you.

Melanie: Meaning?

Sandra: Nothing.

Melanie: I cared.

Sandra: Did you?

Melanie: I shouldn't have come home.

Sandra: For the funeral? I don't believe you said that.

Melanie: I always get crazy when I come back.

Sandra: You always come back <u>knowing</u> everything will be awful, so it is.

Melanie: Sandy, that's not fair.

Sandra: Why can't you just <u>look</u> for a change instead of <u>judging</u>. . . .

Melanie: All I saw was Mom and Dad fighting. Daddy's dead and I still hear them screaming at each other.

Sandra: It's over now.

Melanie: You shouldn't have to hate coming home.

Sandra: I know.
Melanie:
      (Handing the bear to Sandy)
      Sometimes I wonder if maybe I should have stayed so you could have left.

That scene continued as the two women tried to reach common ground, but I suspect the opening sequence is enough for you to see how the student author parceled out information in small segments that lead the audience on instead of giving them everything at once.

How would you rework the chunked parts of the following short passage? Amy and Cassie are sitting at a table in an ice cream shop.

```
Amy: You know, Cassie, I get so depressed when I think
     of this summer. I don't know if I can last
     without Brian. Sometimes I wish I had never met
     him and other times I don't know if I can live
     without him.

Cassie: Wait a minute, Amy. Weren't you just recently
        telling me that you had had it with Brian and
        that you were ready to see someone else?
```

There are several problems in these two lines. First, gobs of information are chunked together. Also, the dialogue is extremely direct. The character's problem arises immediately, and the characters express exactly what is on their minds. Furthermore, they use long speeches with relatively common, clichéd language. Next, the setting isn't used at all. There's no apparent reason to be in an ice cream shop.

Finally, I'm uneasy whenever I hear lines like "Weren't you recently telling me." Beware of any line of dialogue you write that starts with "As I said before" or "You know that" or "I already told you." Such lines usually indicate that the characters already know what's being said, so it's actually only being said for the benefit of the audience. Information that is being rehashed must contribute something new—a discovery of something in the old information or a new angle for a character. Otherwise it's just clumsy exposition.

How did you rewrite the two lines from the ice cream shop? They might be better if they went something like this:

```
Cassie: Good soda, huh?
              (No response)
        Hey, it's summertime. You're supposed to be
        happy.

Amy: I was thinking about Brian.

Cassie: Brian? I thought you'd had it with Brian.

Amy: Well, yeah. I mean, sometimes I wish I'd never met
     him, but other times I don't know if I can live
     without him.
```

There's a better flow to those lines, some sense that the characters are actually paying attention to each other. There's at least some use made of the setting, although there's still no real reason why they're at the ice cream parlor.

The phrase "I don't know if I can live without him" still smacks of cliché. A cliché is a phrase that has been so frequently used that, while the meaning is clear, the words convey little about the character who speaks them. If you find a cliché in your dialogue, examine what the character is actually saying. For example, if a character says "My hands are tied," the actual meaning is, "Someone above me won't let me do anything about it." Reduce the cliché to the basic meaning. Sometimes that basic language will provide the best means of expression. Other times it will guide you to find original words appropriate for a particular character to express the meaning.

Occasionally a cliché might be effective if it fits the way a particular character would talk. Perhaps Amy is unable to express herself in anything except overused language. Playwrights, however, normally try to invest even their most pedestrian characters with means of expression that are a little distinctive. In Tina Howe's *Coastal Disturbances*, the lead character, Holly, expresses a sentiment very similar to Cassie's when she says of a young man she's met: "He makes me crazy, but I'm just so alive with him." If your character must use trite language, perhaps you can devise an interesting way to use it. Amy and Cassie's dialogue might go like this:

Amy: Well, yeah. I mean, sometimes I wish I'd never met him, but other times he's so. . .

Cassie: Divine?

Amy: No. He's so. . .

Cassie: Heavenly?

Amy: No. Neat!

Cassie: Oh.

That makes Amy the queen of triteness.

Writers create natural dialogue and avoid clichés by being careful listeners and observers of the world around them. Just as David Mamet recorded conversations in restaurants, so you, too, should train your ear to hear the peculiarities of speech of the people around you. Then you should write down your observations. Virtually every good writer I know keeps a notebook with him or her at all times to jot down observations, turns of phrase, or ideas. Some writers have drawers full of their old notebooks, and if they hit a block, they turn to those for an idea to get them going again. If you want to be a writer—any kind of writer—keeping a notebook is a practice you should cultivate.

If you aren't already keeping a notebook or journal, perhaps the following exercise, which is designed to increase your sensitivity to the way people express themselves, will get you started.

• • • EXERCISE 4

Select two individuals and listen carefully to the way they talk. Listen for and write down repeated words and phrases, incomplete sentences, particular rhythms and cadences, pauses, and unusual word choices. Look for nonverbal contributions to the conversation such as gestures, looks, and inflections.

Write a scene of conflict in which you incorporate into the dialogue some of the speech habits you observed.

EXAMPLE

GOOD IDEA, IF IT WORKS
by Dwayne Yancey

THE SETTING is a gas station at night. At right is a large front window and the door, which leads out to the pumps. In addition to the weak fluorescent lights overhead, there is a pale light shining through the windows from the station sign outside. There is a pyramid of oil cans and a tire display in the front window. A large side window covers upstage. In the center is a Coke chest, with a SCRUFFY–LOOKING TEENAGED BOY leaning against it. He is dressed in torn and faded blue jeans, an old T–shirt, a denim jacket, and a toboggan–style cap. He holds a bottle of Coke in his hand, his arms crossed, and he is looking at the floor. At left is a display case filled with assorted candies; on top is a cash register and a stack of folded maps. Sitting on a stool behind the register is ANOTHER SCRUFFY–LOOKING TEENAGED BOY, similarly dressed, but without a hat. He is smoking a cigarette. On the wall behind him is a rack of cigarettes, a glass case filled with tapes, various pegs draped with fan belts, a shelf of sparkplugs, and a faded state highway map. Presently the boy at the Coke chest speaks without looking up.

MIKE

So whaddya think?

JEFF

(Looking up, confused)

'Bout what?

MIKE

(Impatiently)

'Bout what we was talkin' about last night.

JEFF

Oh––yeah.

(Pause)

Be all right, if it worked.

(He looks down.)

                              MIKE
                    (With conviction)
          It'll work.
                    (Pause)
          Hey, Jeff.
                    (The boy behind the counter looks up.)
          Gimme a pack of cigarettes.

                              JEFF
                    (He turns and surveys the rack.)
          Whaddya want?

                              MIKE
          I don't care.
                    (Watches as JEFF pulls out a pack)
          Nah, not them.
                    (JEFF's hand moves to another brand. He
                    turns to get MIKE's approval.)
          Yeah, they'll do.
                    (JEFF tosses him the pack.)
          Say your boss never misses 'em?

                              JEFF
          Naw, How could he? He'd never notice.

                              MIKE
          No way he could.

                              JEFF
          That's what I say.

                    (They laugh. MIKE lights his cigarette,
                    walks to the door and looks out.)

                              MIKE
          Boss a pretty good fella?

                              JEFF
          Yeah, he's pretty good. Easy to get along with.

                              MIKE
          That's good.
                    (He smokes, takes a swig of Coke, walks
                    toward the counter, leans across it, and
                    looks at JEFF.)
          How much you reckon ya got in there?

                              JEFF
          I dunno. Couple a hundred dollars. Probably not that
          much. I dunno.

                    MIKE
          (Lets out a low whistle)
What time does your boss come by to lock up?

                    JEFF
'Bout twelve. Sometimes before.

                    MIKE
What time is it now?

                    JEFF
          (Looks at his watch)
Goin' on eleven.

                    MIKE
          (After a long pause, looking at JEFF and
          smiling)
Well, whaddya think?

                    JEFF
I dunno, Mike. It'd be nice all right, but ya think it
would work?

                    MIKE
Now's as good a time as any.

                    JEFF
Yeah.

                    MIKE
Might as well do it now, don't ya think? Wait any later
and your boss might come by to close the place up.

                    JEFF
Yeah, I reckon so.

                    MIKE
          (Walks to the door and looks both ways at
          the road)
Nobody else is gonna be comin' in now.

                    JEFF
No.

                    MIKE
          (Looking at Jeff)
Well, all right then.

                    JEFF
          (Inhales deeply)
Okay. Who do ya wanna say did it?

                    MIKE
We won't say anybody. We'll just make up somebody.

                    JEFF
Yeah, I know that, but what do they look like and all?

                    MIKE
I dunno. What sounds good?

                    JEFF
Wanna say some skinhead done it?

                    MIKE
Yeah. That's good. They'll believe that. That'll be all
right.

                    JEFF
Tall or short?
          (MIKE nods.)
Which one?

                    MIKE
Tall. But not too tall. Nothing unusual. Six foot, six
one. Something around there.

                    JEFF
How about weight?

                    MIKE
Oh, about two hundred. Yeah, that's good. Sounds about
right.

                    JEFF
What should he be wearing?

                    MIKE
I dunno. Just regular clothes. Don't get into too much
detail, or it'll start soundin' funny. Tell 'em you
didn't get a good look at him, it happened so quick and
you was scared.

                    JEFF
Yeah. How about a mask?

                    MIKE
          (Looking out the door)
All right.

                    JEFF
Anything else? How about a car?

                    MIKE
Naw. Then ya gotta start givin' models and colors and
all that crap. Just say he was on foot and leave it at
that.

                         JEFF
Okay. What should I do afterwards? Meet you someplace?

                         MIKE
Shoot no! It'll probably take ya awhile talkin' to 'em
anyway. I'll meet ya someplace tomorrow. I'll go over
to your house.

                         JEFF
Okay.

                         MIKE
          (Turns from door and walks to register)
All right. Gimme a paper bag.
          (JEFF pulls one from beneath the counter and
          shakes it open.)
That looks good.

                         JEFF
Whaddya wanna bag for?

                         MIKE
Might wanna stash it someplace, just in case.

                         JEFF
Oh.

                         MIKE
          (Looks at JEFF)
Go ahead.
          (JEFF hesitates.)
I'll watch out.
          (MIKE returns to the door and looks out.)
It's okay. Go ahead.
          (JEFF takes money out of the register and
          deposits it in the bag.)

                         JEFF
Coins too?

                         MIKE
Yeah, why not.

                         JEFF
          (Finishing)
Okay, I got it.

                         MIKE
          (Takes bag and stuffs it under his coat)
Okay. Call 'em. I'll leave soon as you call.

                    JEFF
            (Looking through phone book)
When you gonna come over tomorrow?

                    MIKE
I dunno. Afternoon probably.
            (JEFF starts to dial.)
Sound kinda scared when you talk to 'em.

                    JEFF

How scared?
            (Voice shaking)
Like this?

                    MIKE
No, not that much. Just choke now and then. It'll sound
all right.

                    JEFF
            (Into phone)
Ah, yeah. I'd like to report a robbery.

            (MIKE smiles, nods at JEFF, waves, and
            quickly EXITS.)

                THE END

EVALUATION

In this subtle battle Mike wants to convince Jeff to rob the gas station, and eventually he is successful. The objectives are clear and in opposition. In one sense the action here is psychological and internal—Jeff's decision to go along with the robbery. But the author also makes the action physical in that Mike both intimidates and encourages Jeff into the decision, which is the crucial moment of the piece. There follows, of course, the outcome, the physical action of getting the money, but that is simply finishing the scene. The conflict has been decided before that even starts.

Virtually all of the dialogue arises from and is focused on the action of the scene. The characters speak because they want things or they have to respond. And their distinct personalities emerge through their interaction. The author of this scene has listened and observed carefully. The dialogue is realistic, yet distinctive. The phrase that sparked the title, "Be all right, if it worked," includes both a statement and a qualifier that makes a kind of crazy sense. The repetition of a word in " 'Bout what we was talkin' about last night" is similarly expressive. The author is also aware that when Mike says "I don't care" which cigarettes he gets, he actually does. He has also incorporated nonverbal cues. For example, Jeff turns to get Mike's approval for the cigarettes without actually saying anything.

The characters have a shared vocabulary, but they are clearly distinguished. Mike is the dominant force. He initiates and directs the conversation about the robbery, the cigarettes, the boss, and the money. Jeff responds only in bits and pieces. Jeff, however, is much more concerned than Mike with the details of the crime: Who should he say did it? What does he look like? What kind of car does he drive? When and where will they meet? Still, even in those concerns Jeff looks to Mike for guidance.

In this scene the author has also noted important aspects of the setting. The old-fashioned Coke chest and faded map indicate that this is not a modern convenience gas stop. The dim interior light and the light from the outside sign suggest a shadowy mood.

Do we care about the characters? Insofar as we have all been subject to petty temptations of greed, probably so. Certainly the author succeeds in getting us interested in whether they will take the money and whether they will get away with it if they do.

The author is also successful in suggesting a future in which the inexperienced Jeff will not see the money again. After all, if Mike fails to come by Jeff's house the next afternoon, Jeff can't really cry "foul" very loudly. If Jeff should weaken and confess, Mike, with the money carefully hidden away, can simply deny any involvement whatsoever.

Now we have seen how any two characters can come into direct conflict. We have examined how characters emerge through action and conflict and how characters express themselves in dialogue that springs from their objectives and the way they go after them. Now let's look at the wild complications that ensue when you add an extra ingredient to the mix—a third character.

5

· · · · · · · · ·

Three-Character
Conflict

EXERCISE 5 • THREE-CHARACTER CONFLICT
Write a scene that places three characters in a situation of modulated conflict. Write the scene using correct stage terminology, with dialogue as needed, and bring the scene to a resolution.

Adding a third character to conflict is like adding a joker to a card game: The possible outcomes multiply dramatically. Mathematics plays a part. Let's say you have two characters/for our purposes, Fred and Madge. Only four basic configurations are possible to express their feelings toward each other.

1. They like each other.
2. They dislike each other.
3. Madge likes Fred, but Fred dislikes Madge.
4. Fred likes Madge, but Madge dislikes Fred.

Fred and Madge, of course, can be unique personalities and their relationship subject to infinite degrees and means of passion, but the relationship eventually resolves itself into one of those four patterns.

When a third character is added (let's say, Hank), the number of possible dramatic structures for the three characters skyrockets. The four relationships listed above are possible between Hank and Fred, and four more are possible between Hank and Madge, in addition to the original four between Fred and Madge. Instead of four possible relationships, you now have four times four times four possible relationships. Sixty-four combinations instead of just four with the simple introduction of a single character!

This point is not intended to make writing a play into an exercise in mathematics. The numbers merely suggest the incredible complications that can ensue when a playwright moves from two characters to three.

We read in theater histories that when Aeschylus wrote the first Greek tragedies some 2500 years ago he used only scenes of direct confrontation between two characters. Sophocles, whose career overlapped that of Aeschylus, is credited with the introduction of a third character into scenes. Perhaps only a playwright can fully appreciate that contribution of Sophocles, for, as we will see, the third character opens a mine of possibilities.

The three-person conflict exercise in this chapter is an extension, but an important extension, of the second exercise. If conflict is basic to drama, then playwrights had better learn how to handle conflict. When two people are involved, the conflict is generally simple and direct: A is against B and B is against A.

As soon as a third party is introduced, however, the situation becomes more complex. Think, for instance, of two teenaged brothers arguing about using the family car. A direct conflict. Then the father walks in. Consider the possibilities. Each boy pleads his case, and then the father decides. Perhaps the father refuses the car to both. Perhaps the boys attempt to cover up their dispute in their father's presence in order to present a united front. Perhaps the father has a special fondness or a particular dislike for one son or the other. A good playwright will be able to use the various possible configurations and shift effectively from one to another.

As the conflict unfolds, the characters will naturally reveal themselves through their actions, their words, their tactics, and their arguments. It is precisely that situation of a father and two sons that Arthur Miller employs so expertly in *Death of a Salesman*. Although his characters aren't just arguing about a car, Miller uses all the strategies mentioned above plus many more in the course of his drama.

As another possible three-character situation, imagine newlyweds having a lovers' spat. Then the young man's mother enters. Again the choices for modulating the conflict between the young couple are extensive. We could envision the mother supporting the daughter-in-law who is trying to straighten out her son. We could imagine the mother supporting her son because he's her child. The boy and girl could unite in telling the meddling mother the fight is none of her business.

As you can see from those hypothetical brothers and newlyweds, the three-person situation allows the playwright certain opportunities that do not exist in a two-person scene. A two-person conflict relies on a direct transaction. That is its great strength. One person desires something and a second person opposes that desire. The scene may contain a panoply of tactics and psychological ploys, but the action always has an essential one-on-one component.

When a third person appears, an important indirect element arises. This example illustrates what I mean. A diamond is on a table and two people are standing in the room with it. Suddenly the lights go out, and when they come on again, the diamond is gone. Because of the one-to-one relationship, both parties in the room must know who took the diamond. The one who took it certainly knows, and the one who didn't take it must know the other person did.

But, you suggest, couldn't someone else have come in when the lights were out and taken the diamond? Aha! You've just discovered the importance of the third character! With three characters in the room, the majority of the characters cannot know with certainty who has the diamond. Only the thief knows for sure. Now suspicion is possible; alliances are possible; deception and pretense are possible; majorities and minorities are possible. In short, conflict that is indirect and variable—or what I call "modulated"—replaces direct conflict as the primary form of interaction.

So important is the three-character situation that certain three-party scenes of modulated conflict have become standard dramatic fare. With infinite variations, they are used over and over again.

A trial is a typical three-party modulated conflict. Two parties modulate their direct conflict by attempting to convince a powerful third party to decide in their favor. That is the essential configuration of the conflict in every courtroom drama ever written. The complainant and the defendant are in opposition, but instead of fighting with each other directly, both are trying to make their case to the third party, represented by the judge or jury. Although more than three characters are usually involved, the conflict is essentially triangular.

The same "trial" structure, which is basically an appeal to authority to make a decision, forms the basis of many other conflicts. Two young men contend for the love of a woman, and they appeal to her to decide between them. Siblings vie for the attention of a parent, each pleading, in effect, "Say you love me more."

A slight variation of the trial format is the peacekeeper. In the example of the trial, two contending parties appeal to the third party to make a decision. The peacekeeper steps between contending parties in an attempt to separate them. The peacekeeper may or may not mediate the conflict, but the immediate goal is to prevent direct conflict. Such is the position of a sheriff who breaks up a fight between two toughs.

The next logical step in the three-sided situation has two of the parties joining forces against the third. The judge hands down a verdict. The girl chooses her beau. The point at which such a decision is made is crucial for a playwright, for once two of the parties (A and B) join forces against the third (C), the situation is for all practical purposes changed to a direct conflict (A/B versus C). Hence good writers try to use the pairing of forces in original ways. Perhaps the audience will expect B to join A, but B joins with C instead. Perhaps A and B oppose C on one issue in the play, but B and C oppose A on another issue, and A and C oppose B on yet a third issue. Through creative pairing of forces, a playwright can produce a rich interplay of shifting alliances.

Another common three-sided situation is a circular one. A wants to gain B's support in opposing C, while B wants to gain C's support in opposing A, and C wants to gain A's support in opposing B. Or the same circle can be created in terms of desire. A wants B, who wants C, who wants A. That circle of desire joined with an equal circle of hate comprises the classic "hellish" situation created by Jean Paul Sartre in *No Exit*.

Another frequently used three-way situation develops when at least one character is not aware of the presence of a third party. Such is the case when one character is hidden or is listening in on a conversation. That arrangement has produced some of the most famous scenes in all of drama, including the hilarious "china scene" in William Wycherly's *The Country Wife* and Hamlet's emotional scenes with Ophelia and Gertrude, when Polonius is an unseen listener.

A scene that is closely allied to the hidden character is one of disguise, in which at least one of the characters does not know the true identity of another character. Even if a scene of mistaken identity contains only two characters, it assumes a third character. If I think you are Elvis Presley returned to life, then our scene contains you, me, and the assumption of Elvis. Such mistaken identity scenes have been a cornerstone of comedy at least since Plautus wrote *The Twin Menaechmi*, in which identical twins are regularly confused for each other. Shakespeare used the same idea in *The Comedy of Errors*. More recently, Larry Shue used similar tactics to propel his comic hits *The Foreigner* and *The Nerd*.

Scenes of hidden and disguised characters are two examples of what is called "dramatic irony." Dramatic irony occurs when the audience knows something that is unknown to a character. The audience, for example, knows someone is hiding in the "china" scene and the *Hamlet* scenes, although some of the characters in the play do not. The audience knows there is a set of twins in the Plautus play and two sets in Shakespeare's comedy, although none of the characters realize it.

Yet another standard three-party conflict is that in which two characters are aware of the presence of a third but are trying to hold a private conversation and avoid being overheard. That format differs from the situation of dramatic irony in which a hidden character overhears a conversation. In this instance all the characters are aware of each other's presence, so there is no dramatic irony, but two of the characters attempt to exclude the third. An example might be a pair of secret lovers in a restaurant trying to keep their conversation from an all-too-attentive waiter.

Finally, there is the "love triangle." A love triangle does not refer to one distinct relationship among three people. Rather, there are numerous variations, all of which qualify under the rubric "love triangle." I have already mentioned one love triangle in discussing the "trial" format—two men both love the same woman, and they appeal to her to make a decision—and another in the circular format where A loves B who loves C who loves A.

Another love triangle could involve a husband or wife who loves his or her mate, but the mate loves someone else. You could have someone who is loved by two individuals who may or may not know each other. The possibilities are numerous enough to have allowed inventive playwrights over several centuries some measure of fame and fortune. Many modern playwrights have given new life to the standard triangles by altering the usual gender of the participants. The essential triangular love relationships, however, remain the same.

As you undertake the three-person exercise described in this chapter, remember that you can learn not only from what *you* write, but from what *others* write, as well. If you are in a classroom situation, pay careful attention to the differences in the three-person scenes. Consider how the relationships between the three characters could be altered or rearranged. If you are not in a playwriting class, give particular consideration to three-person scenes in the plays, movies, and television shows you see.

Once a writer has learned to handle competently two characters in a direct conflict and three characters in a modulated conflict, he or she can handle virtually anything. More characters or new alliances merely present variations based on direct or modulated conflict. As an example of what I mean, recall the newlyweds who were arguing. Let's assume they're fighting because he wants to play softball, and she wants him to do more around the house. Assume both of the young girl's parents enter. Assume further that the mother never liked the young man her daughter married, and she sides with the daughter. The father didn't like the young man either, but, after the boy explains his position, the father sides with him because he feels down deep that his own wife always prevented him from doing things he liked to do. Now let's examine how such a scene might be structured.

Scene Action	*Scene Structure*
1. The young man (A) and young woman (B) argue.	1. Two-party direct conflict: A vs. B
2. The woman's parents (C and D) enter. The daughter and her husband appeal to them for support.	2. Three-party modulated conflict ("trial" format): A vs. B, both appealing to C/D
3. Mother and father both side with daughter.	3. Two-party direct conflict: A vs. B/C/D
4. Young man admits errors, but appeals to father-in-law's sporting sense and sense of fair play. Father attempts to mediate dispute and prevent further conflict.	4. Three-party modulated conflict (father as "peacekeeper"): D tries to mediate between A vs. B/C
5. Father supports son-in-law's desire to play softball.	5. Two-party direct conflict: A/D vs. B/C
6. Mother attacks father for not supporting daughter; father reveals resentment toward mother.	6. Two-party direct conflict: C vs. D
7. Young man and woman join in attempt to end strife between mother and father.	7. Three-party modulated conflict (young couple as "peacekeeper"): A/B try to step in between C vs. D
8. Mother and father storm out arguing; young couple resolve to compromise.	8. The scene, even if some problems remain, is resolved

The two-party conflict that begins the scene ends as the young husband and wife join forces to stop the fight between the woman's parents and then determine to resolve their own differences. Notice that at no time is there actually a four-way conflict. I would not go so far as to say that a four-way conflict is impossible, but I would assert that, as parties proliferate, they usually team up to create combinations of direct two-party or modulated three-party conflicts.

• • • EXERCISE 5

Place three characters in a situation of modulated conflict. Use stage terminology with dialogue as needed to write the scene, and bring the scene to a resolution. The scene should be at least five pages.

The factor that distinguishes this exercise from the conflict exercise in Chapter 2 is, naturally, the presence of a third character. In evaluating the scenes you write, ask yourself some crucial questions: How does the presence of three people affect the conflict? Can you identify what each character wants and how he or she goes about getting it?

Keep in mind also the fundamentals of the previous exercises. What action does the audience see taking place? Does the dialogue flow naturally from one character to another? What kind of language does each character use?

This exercise can be done in two ways. You should write a completely new scene with three characters. After you have done that, however, you may wish to add a third character to one of your previous two-person scenes to see how that changes the complexion of the scene.

To understand how the addition of a third character can affect a scene, let's examine a student-written piece that begins with two characters and then adds a third.

EXAMPLE

BREAK
by Debbie Laumand

(MOM is seated in a chair in an ordinary living room. She is reading a book. BETH, age 21, ENTERS. She is carrying a book.)

BETH
I thought I'd find you in here reading.

MOM
(Not paying attention; keeps reading)
Yes, dear.

BETH
I like to read in here. It's nice and quiet.

MOM
(Nods head vaguely)
Um-hum.

                        BETH
Nobody to bother you. You know what I mean?
                    (MOM nods again. BETH makes a horrid face at
                    her mother to see if she's paying attention.
                    She isn't.)
I'm gonna read some Shakespeare, Mom.

                        MOM
                    (Nodding)
Um-hum.

                        BETH
You know what my English professor says about
Shakespeare?

                        MOM

Um-hum.

                        BETH
He says if I don't read Shakespeare, I'll become frigid
before I'm twenty-five.

                        MOM

That's nice, dear.

                        BETH
Yep, Mom. That's right. The ol' "spread the legs and a
little light turns on" routine. No Wil, no will. You
know what I mean? Ice cubes, Mom. I'm talking crushed
ice.

                        MOM

That's nice dear.

                        BETH
                    (Sighs)
I guess I'll read now.
                    (BETH sits and begins to read. A door
                    opening and closing is heard offstage and a
                    light infiltrates, as if someone has turned
                    on a light in another room.)
I guess Ben's home.

                        MOM

Um-hum.

                    (A TV set is heard from the next room. Loud.
                    A football game is on.)

                        BETH
God! There go the holiday football games. Does he turn
that dumb thing on the second he walks in the door?

(No reaction from MOM. BETH gets up and goes
to the entranceway. She yells at BEN.)

Must you subject Mother and myself to that mindless
drone at decibel rates far beyond the endurance of the
human ear? Is it absolutely necessary to have the sound
so loud that one's hair is parted by the sheer force of
it? You may not understand this, Ben, but mother and I
are practicing an ancient art called "reading." We are
expanding our minds with the written word as opposed to
shrinking it with an array of unconnected molecules,
more commonly known as the "television image." And we
don't appreciate your blatant and rude behavior in
turning. . .

> BEN (Off)

Aw, shut up, Beth. You're fulla shit.

> BETH

You watch your language. You may be fifteen now, but
that doesn't mean you can speak to me that way.

> BEN (Off)

You're fulla crap.

> BETH

I am merely trying to express my thoughts in the most
civilized manner possible. I am trying to reach you
through the spoken language. I am asking you nicely to
turn the volume down.

> BEN (Off)

Go to hell!

> BETH

You turn that volume down or I'm gonna come in there
and beat the crap out of you, and I'm still big enough
to do it, bucko!

(The volume is turned down. MOM looks up at
BETH. BETH catches the look. MOM looks down
and resumes reading. BETH crosses to her
chair and picks up her book.)

The only language these heathens understand is
violence. I don't believe him. He didn't act like that
when I lived here, did he? I mean, what has happened to
this family?

(No response. BETH begins to read again.
Suddenly BEN is heard screaming rapidly from
the other room.)

                    BEN (Off)
All right. All right! He's got the ball! Run, you
sucker run! Do it! Do it, man! Do it! Do it! Do it!
Yeah! Yeah! Yeah! Whoa!

                    BETH
              (Starting, eyes wide)
Good God! What is he screaming about? Does he always do
that?

                    MOM
              (Not looking up)
Yes, dear.

                    BETH
Jeez, my heart's pounding like crazy.
              (No response from MOM. BETH returns to her
              book.)

                    BEN (Off)
What the . . . ! What kinda ref are you? Are you nuts?
Aw, what a bummer!

                    BETH
              (Starts, looks up, and yells)
Knock it off, Ben!

                    BEN (Off, loud)
Oh, man! What a catch! Go with it, man. All right! Go!
Go! Do it! Do it! Do it! All right! Wow!

                    BETH
              (Looks down at book and overlaps LOUD as BEN
              continues to yell)
Oh, man! What a bummer! The poison won't kill Juliet.
But wait. She's going for the knife! All right! She's
gonna stab herself. All right! Do it! Do it do it do it
do it! Yeah! All right! Way to go, Juliet!

                    BEN
              (At entranceway)
Beth, you're a real hag, you know that?

                    BETH
              (Goes to him)
Listen, twerp, I've had just about enough of you!

                    BEN
Yeah? Well why don't you do something about it?

                         BETH
All right jerko . . . you better believe I'm. . .

                         MOM
              (A forceful statement, but not a yell)
That will be quite enough, you two.

                         BETH
But he. . .

                         MOM
Beth, quiet. Ben, you watch your language and stop
provoking your sister.

                         BEN
Yes, ma'am.
              (BEN RETURNS TO THE OTHER ROOM.)

                         BETH
Good. You really have to keep a strong control over
these kids or they just. . .

                         MOM
And you, young lady.

                         BETH
Ma'am?

                         MOM
You stop provoking your brother.

                         BETH
He was the one with the screaming problem.

                         MOM
Beth, he lives in this house just like you do, and he
has a right to be here, just like you do.

                         BETH
But he was disturbing my reading. How could I. . .

                         MOM
My! My! My! Me! Me! Me! You've been a tyrant since day
one of your vacation.

                         BETH
I have not. . . .

                         MOM
Yes you have. I don't know what you're going through
right now, Beth, but it has got to stop. You were
treating your brother as if he didn't belong here.

                         BETH
Everybody's changed. I don't feel like this is my house
anymore. I was so looking forward to the Christmas
break, Mom. I just wanted it all to be like . . . well,
like a Hallmark card.

                         MOM
Beth, this is your home. It's the same home it's always
been, and we love you. And if you don't straighten up,
"I'm gonna beat the crap out of you, and I'm still big
enough to do it, bucko!"

                         BETH
              (Laughs)
I'm sorry, Mom.

                         MOM
That's ok. I'm going into the kitchen to get myself a
glass of tea. Do you want any?

                         BETH
Yeah. I'll go with you.

                         MOM
Good. But leave your Shakespeare book here, darling.
The freezer's broken and I want to see you make some
crushed ice.

                         BETH
Mom!

              (THEY EXIT.)

              THE END

EVALUATION

The basic conflict of this scene is between a young woman who
wants attention and her mother, who thinks her daughter is too self-cen-
tered. The playwright has an inherent problem to solve. Since the mother
is ignoring the daughter, the scene seems to have nowhere to go. Until
the introduction of the third character. The arrival of Ben in the next
room provides Beth with an opening, an excuse to create a disturbance.

At its core, the scene has the structure of a "trial" situation. Two
siblings are contending for the approval of a parent. What makes the
scene unusual is that the conflict between the siblings is deemphasized in
order to increase the attention on the relationship between the girl and
her mother.

The writer makes an intriguing choice in keeping the third character almost entirely offstage. Normally that would not be a wise choice,
and the image of the girl standing at the doorway talking to an offstage
character might seem contrived. How would it have affected the scene if
Ben were on stage more? Naturally Ben would have become a much more

prominent figure. And the direct conflict between Beth and Ben would have been the central element of the scene. By keeping Ben offstage, the focus remains on Beth and her mother.

Are the characters differentiated in the way they speak? Beth conveys several qualities through her language. Her speech has a broken, jagged quality that suggests youth and quickness: "No Wil, no will. You know what I mean? Ice cubes, Mom. I'm talking crushed ice." Her lines reveal not only that she is educated, but that she is bright and witty. That information is not conveyed through direct means. No one, thank heavens, says, "My what a bright and witty girl you are." We don't conclude she's intelligent because she's in college or because she's reading Shakespeare. Lots of not particularly bright individuals go to college and some even read Shakespeare. What convinces us that Beth has a good mind is in part her ability to handle language: "Must you subject Mother and myself to that mindless drone at decibel rates far beyond the endurance of the human ear?" But even more, we are convinced of her intelligence by her sense of style, by her ability to shift from overblown verbiage to common language in a trice, as when she moves from "I am merely trying to express my thoughts in the most civilized manner possible. . . ." to "You turn that volume down or I'm gonna come in there and beat the crap out of you. . . ." in the space of a single line.

Beth's consciously elaborate language and her mocking imitation of her brother's speech reveals that she is self-dramatizing, and in the drama that she creates she is also quite entertaining. That, after all, is a must for plays.

The audience finds out much less about the other two characters through the language they use. Ben talks in slang at virtually every line: "What a bummer" and "You're a real hag." From that we get a sense of what Beth is attempting to place herself above. Still, through her calculated use of similar slang we understand that it remains a basic part of her.

The mother has few lines, yet they disclose her character fully. Her repetition of "Um-hum" and "That's nice, dear" indicates someone who is not paying attention and suggests that Mom is not particularly interested in her daughter. We soon find, however, that Mom is in charge in her home. She effectively stifles the spat between Beth and Ben, and her direct, plain talk to Beth shows her as both forthright and concerned. Finally, her last lines demonstrate that she heard and cared about everything her daughter said.

Now let's look at a tight, complex scene that uses three characters throughout.

EXAMPLE

NIBBLES

by Mary Kerr

THE SETTING is a living room. There is a couch center, facing the audience. There is a small portable TV on a low table down center in front of the couch. MIKE and DREW sit on opposite ends of the couch watching TV. Both men are about 20. STEPHANIE, about the same age, ENTERS carrying a bowl of chocolate chip cookies. She sits between the two men.

STEPHANIE

I made some cookies to eat while we watch the game. They're still hot.

MIKE

All right!
>   (He grabs a handful of cookies out of the
>   bowl and begins eating.)

STEPHANIE
>   (Offering bowl)

Drew?

DREW

Oh, no thanks.

STEPHANIE

You're sure?

DREW

Yeah. Thanks anyway.

STEPHANIE

Come on. They're real good.

DREW

I don't want any, Steph, really.

STEPHANIE

Here I go to all this trouble, and you're not going to eat any?

MIKE

Drew's on a diet.

STEPHANIE

What? Drew, is that true?

DREW

Yeah.

STEPHANIE

But that's silly. You're not fat.

DREW

I'd like to drop about five pounds.

                    STEPHANIE
That's crazy. You look fine. Come on, have a cookie.

                    MIKE
He doesn't want one.

                    STEPHANIE
One cookie won't hurt him, Michael.

                    MIKE
Look, Steph, just leave him alone, ok?
          (MIKE takes four more cookies from the
          bowl.)

                    STEPHANIE
Don't eat 'em all. Save some for Drew.

                    DREW
I really don't want any.

                    STEPHANIE
But you aren't fat!

                    MIKE
Geez, the guy is a little overweight and he's trying to
do something about it and you won't leave him alone.

                    DREW
Hey, yesterday you said you didn't think I needed to
lose any weight.

                    MIKE
Well . . . ah . . . I just meant that you thought you
were overweight. I don't think you're overweight.

                    STEPHANIE
Neither do I, Drew. So have a cookie.

                    DREW
Don't tempt me, Steph. I really don't want one.

                    MIKE
Good! That's more for me.
          (He reaches into the bowl for more cookies.)

                    STEPHANIE
Mike, you could afford to forgo a few cookies yourself.

                    MIKE
Oh, really?

                    STEPHANIE
Yes, really. You're sprouting some pretty hefty love
handles.

                         MIKE
Yeah? What about you? Those jeans weren't that tight
when you bought 'em.

                      STEPHANIE
Am I getting fat?

                         MIKE
I didn't say fat. Your butt's just gettin' a little
wider is all.

                         DREW
Don't listen to him, Steph. Your butt looks just fine
to me.

                         MIKE
When were you lookin' at her butt?

                         DREW
Jeez, Mike, she lives here. I see her butt every day.
It's not like I've been staring at it.

                      STEPHANIE
          (Standing up, looking down at her rear)
Does it really look alright?

                         DREW
It looks fine. Your whole body looks fine.

                      STEPHANIE
So does yours.

                         DREW
Thanks.
          (STEPHANIE and DREW smile at each other as
          she sits back down and they resume watching
          TV.)

                         MIKE
There's three cookies left. Drew?

                         DREW
No thanks.

                         MIKE
Steph?

                      STEPHANIE
No thanks. I think I've had enough. Go ahead, Mike, you
finish them off.

                         MIKE
No, uh, I'm not hungry anymore. We can save 'em for
Kurt for when he gets off.

                    STEPHANIE
That's a good idea. Kurt can eat them.
                    (They sit looking at the TV in silence.
                    STEPHANIE lights up a cigarette. MIKE gets
                    out a piece of sugarless gum. DREW begins to
                    chew his fingernails.)

                    THE END

EVALUATION

On purely technical grounds this scene is an excellent example of what can be accomplished with a three-person scene built on shifting alliances. It begins with Stephanie (A) thrusting cookies upon Drew (B), who refuses (A versus B). Mike (C) supports Drew's refusal (A versus B/C). The conversation moves to the topic of weight. Drew maintains he's overweight and Stephanie disagrees. At first Mike seems to agree with Drew, but in one of the nice ironies of the scene Mike is forced by the person whose side he is on to switch sides ("I just meant that you thought you were overweight. I don't think you're overweight.") Mike's switch thereby generates A/C versus B. Stephanie immediately upsets that arrangement by insisting that Mike is overweight (A versus C), and he retaliates by accusing Stephanie of heftiness. Drew, however, supports Stephanie, creating an A/B versus C situation and bringing together the two people who were at odds at the beginning of the scene.

The scene is notable in a number of other ways. First, despite its limited physical action it contains large psychological movements. Second, it produces those movements in a spare, concise manner. Third, it conveys a strong sense of sexual tension without any direct reference or overt symbolism.

The physical action of the scene is limited to three young people watching television and eating cookies. The conflict at the surface of the scene is Stephanie's desire for Drew to eat the cookies she's just made and his desire, supported by Mike, not to eat any cookies.

That leads to a second level of conflict. Is Drew fat? Drew thinks so, Stephanie doesn't, and Mike's position isn't clear. Is Stephanie fat? Mike says so, but Stephanie and Drew say otherwise. Is Mike fat? Stephanie suggests he is, but Mike denies it.

Physically the scene is nearly static, with only the passing of the bowl and the munching of cookies. But psychologically the ground covered is vast. Mike goes from ease and confidence to insecurity and virtual denial of appetite. He begins as a friend of both Stephanie and Drew, but he winds up strangely alone. Drew shifts from alliance with Mike over his own need to lose weight to alliance with Stephanie. Stephanie, in effect, moves from Mike to Drew, and we see in small details the end of one romantic relationship and the beginning of another.

Psychologically, then, this scene is about sexual relationships, and it is to the author's credit that she has written a highly charged, erotic scene with hardly a word of sexually explicit language. Yet the sexual overtones are clearly there, in the focus on Drew's body, in the emphasis on Stephanie's "butt," in the reference to Mike's "love handles," and in every gluttonous mention of the chocolate chip cookies.

Although there is much in this scene to admire, there are also ways in which it might be improved. For one, we completely lose sight of the fact that they are watching television. Since that is, on the surface, a primary action, it ought to be used. The scene would benefit from references to appropriate images on the screen—perhaps weight-lifters, body builders, or 300-pound tackles. As the scene now stands they might as well be doing a crossword puzzle, and, in fact, perhaps there would be a better activity for them to be engaged in than watching television. By "a better activity" I mean one carefully but discretely connected to the psychological action of the scene.

A second primary action of the scene is eating the cookies, and there, too, more could be done with the action. Some attention might productively be drawn to the ways that Stephanie and Mike eat their cookies. Perhaps Mike makes a mess, getting crumbs on his shirt or dropping pieces of cookie on the sofa. Perhaps Stephanie eats her cookies slowly, nibbling around the chocolate chips, which become sticky on her fingers so that she licks off the wet chocolate. Precisely how to incorporate watching the television and eating the cookies more noticeably into the action is for the playwright to determine, but attention to those primary actions could add scope to the scene.

A second area in which the scene could be improved involves the use of language. The dialogue of the scene flows easily from one character to another, and it sounds natural. We can believe that people talk like that, and that is a solid virtue. What is lacking is clearer differentiation in language patterns between Mike, Drew, and Stephanie. In terms of the words used, and the way the words are put together, any of the lines could be spoken by any of the three characters.

"Nibbles" depends on there being three characters. Without Drew in the scene, Stephanie and Mike would simply have eaten the cookies without a word about weight and without conflict. Without Mike, Drew would simply have refused the cookies and would never have declared how appealing he thinks Stephanie looks. And, of course, without Stephanie and the cookies there is no scene. Such are the complications of the three-character scene.

Have you developed a sense for creating action? For showing your audience characters doing something? Can you put two characters into a direct conflict and manage the scene? Have you gained a sense of how people express themselves in language? Have you come to terms with the variety of ways three parties to a conflict can interact? If so, then you have the basic tools needed to prepare any scene in drama. In the following chapters you'll be able to use those tools to develop additional scenes that will bring your characters and your ideas to life.

6
• • • • • • • •

Writing from Life

EXERCISE 6 • WRITING FROM LIFE
Write a scene based on something you know well from actual
life. Use stage terminology and as much dialogue and as many
characters as you need.

In many writing classes—not only playwriting but other forms of
creative writing as well—students are instructed to write from their personal experiences. That is an excellent place to start, and in this chapter
you will be asked to write a scene about something you know extremely
well, a scene that uses as its seed something taken from your own
experience.

All writers write from their own experiences. In some cases the
dramatic creations are quite close to real life events in a writer's life. Tennessee Williams used remembrances of his mother and invalid sister to
help him shape Amanda Wingfield and Laura in *The Glass Menagerie*. Robert Harling was recalling his sister's struggles with diabetes when he composed *Steel Magnolias*. Ntozake Shange, who wrote *For Colored Girls Who
Have Considered Suicide/When the Rainbow Is Enuf*, has stated very candidly
that everything she writes is founded in her race and her gender. Alfred
Uhry used memories of his own relatives in writing the delicate scenes of
Driving Miss Daisy.

Eugene O'Neill's *A Long Day's Journey Into Night* examines family
relationships through characters that mirror O'Neill's own family: a father
who is a famous actor, a mother living in a drug-hazy past, a wastrel
brother, and the playwright's own unhealthy self. Even in his earlier
plays O'Neill's realistic seamen were based on men he had known while
he worked on seagoing cargo vessels.

Tina Howe's early experiences at the Metropolitan Museum of Art
influenced her writing of *Museum* and *Painting Churches*. Her own parents
bear a strong resemblance to the parents in the latter play.

Even when the subject appears to be distant from an author's actual life, the author is still using personal experience. Kings and queens
are frequent subjects of plays. Maxwell Anderson's *Mary of Scotland*, Robert Bolt's *A Man for All Seasons*, Caryl Churchill's *Top Girls*, and James
Goldman's *A Lion in Winter* are just four examples. All such factual materials must be filtered through the eyes and ears of the playwright. Conse-
quently, Anderson presents Mary and Elizabeth in conflict as he has

observed women fighting. For example, Mary taunts Elizabeth with the fact that she has a son and Elizabeth has no children. Churchill's historical characters reflect on women's issues with a twentieth-century consciousness. In *A Man for All Seasons*, Sir Thomas More's family can behave only as Bolt understands family members to behave. In his moment of greatest weakness More is supported by his wife, not because she understands him, but because she loves him *despite* her not understanding him. And Henry and Eleanor in *A Lion in Winter* must reflect Goldman's observations of the struggles of husbands and wives, including the ways they use children, and property, and other romantic relationships to direct their feelings at their partners.

In short, writers write from their own experience because they can't *not* write from their own experience. Who else's experience can they use? Who else's can you use?

Sometimes, particularly in dealing with historical material, a writer conscientiously attempts to use only documented facts. That can lead to good history and biography, but it usually leads to bad plays. I don't mean that facts must be changed or distorted—although Shakespeare and Marlowe certainly had no hesitation about altering history to create better drama. But when you write a dramatic character, that character must come to life on stage. The character must act, talk, move, think. The dramatist cannot possibly know every detail about even a well-known historical figure, and so, as a result, the writer must fill in the details in order to bring that personage to life. And those details must come from the writer's own background and be filtered through the writer's mind and feelings.

I am not suggesting that historical research has no part in the creation of such plays. I'm sure that Anderson and Churchill and Bolt and Goldman studied their subjects carefully. What I am saying is that the research itself becomes part of the writer's experience—an ancient account, if you will, sifted through a twentieth-century consciousness—and that everything with which a person comes in contact, from personal observations to book-learning, becomes a part of that individual's experience.

Writing from your own experience has several advantages. Because you are writing about places, people, and experiences that you know well, the writing frequently contains details that imbue the scene with depth and richness and give to the characters a set of very human idiosyncrasies.

Another advantage in writing about something you know well is that you will usually have some emotional connection with the material. You may even care very deeply about a person or a situation. If you can find a dramatic way to make the audience care deeply as well, you will have accomplished wonders.

This point of caring about your material can hardly be overstressed. In fact, many writers view it as a crucial starting point. Wendy Wasserstein once said that what drives her to write comes from experiences she's actually had as a woman. You must care, and care deeply about your material and what you want to show us if you hope to make anyone else care about it. After all, if you don't care about what you're writing, why should anyone else?

That's why personal experiences often make such a good spring-board to writing. Whether you're writing about a situation you've known or a person who has crossed your life, the chances are you will have a deep personal and emotional involvement. If the situation involves a family member, you will probably have strong feelings one way or another, just as O'Neill did and just as Robert Harling did. Possibly there will even be conflicting feelings of love and hate. All of that can provide fertile ground for dramatic possibilities.

Just as there are advantages to writing from your own experience, there are also potential hazards. One danger lies in an overreliance on an actual situation. I recall one instance when a young woman wrote a very good play about a young woman breaking away from her family and learning to stand on her own. There were several excellent scenes that involved the family. They showed that the family members loved each other but unconsciously stifled each other as well.

A flashy boy, who was essentially a negative character, was instrumental in the young girl's break from her family and its traditions. Most of the audience who heard a reading of the play thought the flashy boy was depicted as too evil and conniving. It did not seem dramatically plausible that the reasonably intelligent girl would have been taken in by him.

The author's response was: "That's how he was." That's the danger. Something may happen in real life, a person may behave in a particular way, and yet when those events are placed on a stage, they may not seem believable. It is inadequate for an author to say "That's how it really was" because, finally, the audience won't know or care how it really was. They will only know what they see on the stage, and that dramatic representation must seem plausible for them to accept the characters and their actions. It is not always possible to predict what an audience will find plausible and what it will reject. Generally, dramatic plausibility means that the audience can accept that the characters could feel the way they are shown to feel or behave the way they are shown to behave.

In *Othello* Shakespeare presents the audience with a strong, active military man. He is a Moor in a white society. He has a white wife whose father is opposed to their marriage. It is plausible that, like many a newly married man, Othello may fear the loss of his beautiful wife's affections. When he begins to grow jealous, he tests her—a plausible action. When his reasons for suspicion and jealousy are apparently confirmed, it becomes plausible that this strong man could, out a sense of betrayal, destroy the thing he loves most in the world.

The playwright, then, must never forget that the responsibility is to the play. The playwright is not a journalist who must recount each detail correctly. Rather, the playwright is an explorer, examining why people act as they do and creating fictional, dramatic figures that must seem right and plausible. As you write, you might keep in mind what Tina Howe likes to say about her plays: "All the events are true, but none of them ever happened."

A second hazard, somewhat related to the first, is that of becoming overindulgent with your own interests. When you know something very well, it is easy to give too much detail about a room, a character, or an event. Only those elements that are dramatically necessary to the action or to the characters need to be included. It is also easy, because you know the situation so well, to provide details that, to you, have obvious connections, but which seem extraneous to an audience.

What happens if you dramatize something that you think is incredibly interesting, but the audience is bored? It's not the audience's fault. You, as the playwright, have failed in some way. Perhaps you have not made clear to the audience the connections you perceive. Perhaps you have simply failed to translate what you found interesting in those connections into something that is dramatically interesting in the world of the play.

Yet another danger of writing based on personal experiences is that of violating other people's privacy. Friends or family members who are involved in an incident that you have used as the basis for a dramatic scene may recognize themselves and not be entirely pleased. Worse yet, other people may recognize them. Bear in mind that when you use personal material as the basis for a scene you are not attempting to recreate reality. You are attempting to construct an entertaining, moving, and perhaps meaningful and instructive dramatic work. To that end, it is advisable to provide some distance by changing names, dates, places, and so forth. More importantly, it might be dramatically necessary to alter events, characters, and motivations to satisfy the particular needs of your dramatic work.

The final danger of writing based on personal experience is a psychological one. Sometimes those things in our lives that are inherently the most dramatic are also the most threatening or the most disturbing. When Marsha Norman wrote her Pulitzer Prize–winning play *'night Mother* about a woman who tells her mother she is going to commit suicide, she said that it was a highly personal piece. All writers must enter such terrain with caution. On the positive side, writing from personal experience frequently allows you to reflect on situations and to understand better both the situation and your own personality and behavior. It can, in other words, help you to "know yourself" and, in so doing, make you a more honest and sensitive writer and person.

Nevertheless, that very process of self-discovery can also be difficult, even painful. If, in examining material to write about, you discover areas that are troubling, you have three choices. One is to write about something else. A second is to pursue the area, while remaining aware that it may cause anxiety, stress, or even more serious problems. I don't wish to sensationalize, but a young writer should be aware that such exploration has led both to excellent plays and to some tortured, damaged lives. A third choice is to seek professional counseling for the problem. That may or may not improve your playwriting, but it might help your adjustment to life, which could be most beneficial anyway.

A final question about using this exercise with young writers: How can young people, who by and large have not had wide-ranging personal experiences, have adequate resources on which to draw? That is really not so great a problem as it seems. True, the experiences of many young writers are limited to those of a student in a classroom and a child in a family. True as well that a group of students using this exercise will more than likely turn up a fair share of plays about school life. But that's all right. Material does not have to be original to be treated originally. In fact, the best material is usually something with which everyone is familiar. There are thousands of plays about family life, about young people coming of age, and about young love. It isn't important that a student write about something that has never been written about before. The student who tries that will probably give up writing in despair. What *is* important is that each student discover a unique viewpoint, a special way of analyzing, understanding, and dramatizing a situation.

Furthermore, don't underestimate the breadth of experience of even a relatively young person. Family and school situations comprise two of the most obvious starting places, but there are numerous other sources: friends, travel, camps, sports, other extracurricular activities, unusual incidents, and jobs are only a few.

With those explanations and warnings, let's look at the exercise and two student-written examples.

• • • **EXERCISE 6**
Write a scene based on something you know well from actual life. Use stage terminology and as much dialogue and as many characters as you need. Bring the scene to a resolution. Five to six page minimum.

You may find as you work within this format that you have a lot to write about. Because you know the situation well, you will have a great deal of material from which to draw. You could even find yourself writing what you think is too much. That's all right. Don't worry about it. Just get down on paper everything you want to write. You can be more selective, if that's needed, in revisions.

Be aware also of the elements you've worked on in previous exercises. What actions do we see occurring? What are the conflicts? Is it direct, between just two people, or is it modulated through the presence of one or more other characters?

The following scene used an incident from the author's experience as a springboard to a scene of direct conflict.

EXAMPLE

HIPPO AND THE BAT
by Mont DeWolfe

THE SETTING is the living room of a small
apartment. It's furnished in a second-hand
manner, with a couch, two easy chairs, a
coffee table and two desks. The items look
very nice individually but fit awkwardly
together. A door to the bedroom is downstage
right, and a light can be seen coming
through the doorway. A doorway to the rest
of the apartment is upstage center, and
upstage left is the entrance to the
apartment.

The door to the apartment opens quietly and
A MAN STUMBLES IN, trying unsuccessfully to
be quiet. A WOMAN ENTERS from the bedroom
and turns on a light. She is tall and
slender with strawberry blond hair, about 25
years old. She is wearing a T-shirt and
underwear.

The man tries to adjust his eyes to the
light. He is tall and lanky, a couple of
years older than the woman. He is wearing
expensive modern clothing, sloppily. He has
long dark hair, and on the left side of his
head is a large clump about the size of a
racquetball. He is drunk and looks guilty.

REBECCA

Dammit, Hippo, where have you been?
      (HIPPO doesn't answer.)
Drunk again, aren't you?
      (HIPPO doesn't answer.)
What'd you do to your head?

HIPPO

Nothing. Where's the scissors?

REBECCA

What'd you do to your head?
      (She moves closer to get a better look at
      his head, recognizes the cause of the clump,
      shrieks, and backs away.)
Hippo, you've got a bat in your hair! Get the scissors!

                    HIPPO
Where are they?

                    REBECCA
I don't know.
          (HIPPO begins searching for the scissors in
          unlikely places such as under the couch.)
Wait a minute. You can't cut it out in here.

                    HIPPO
Why not?

                    REBECCA
It'll attack me.

                    HIPPO
It's dead.

                    REBECCA
Is this a joke, Hippo?
          (HIPPO continues his search, not answering.)
Hippo, is this some kind of lousy joke?

                    HIPPO
It's not a joke.

                    REBECCA
You stand me up, you get drunk--which you promised you
wouldn't anymore--and then you play this sick joke on
me.

                    HIPPO
It's not a joke, Becky. The bat attacked my head.

                    REBECCA
Then how did it die?

                    HIPPO
I strangled it.
          (Finds the scissors in a desk drawer)
Here they are! Why'd you put the scissors in my desk?

                    REBECCA
I don't know. Maybe you used 'em last.

                    HIPPO
Naw, you were using them the other day to cut out that
magazine stuff.

                    REBECCA
Hippo, how did the bat get in your hair?

                    HIPPO
          (Cutting the bat out of his hair in front of
          a mirror)
I don't know. It just swooped down and got tangled in
my hair. So there it was on the top of my head,
shrieking. I was walking along, and then "Bam." At
first I tried to run, then I stopped and sort of
wrestled with it. Then I kinda lost my sense of where I
was and stumbled out into the road and this car hit me.

                    REBECCA
You got hit by a car? Are you okay?

                    HIPPO
Yeah. It wasn't anything serious.

                    REBECCA
Hippo, anytime you get hit by a car it's serious.

                    HIPPO
No. It just grazed me. Anyway, I managed to kill the
bat.
          (The bat finally comes free from HIPPO's
          head and falls to the floor.)
There. Thank God.

                    REBECCA
Did the car stop?

                    HIPPO
Yeah. I think he was more shaken up than I was. So he
gave me a ride home.

                    REBECCA
          (Pulls HIPPO away from where the bat is
          lying on the floor and inspects his head.)
It doesn't look like it bit you. I don't see any blood.
          (She lets go of him.)
I'm still mad at you, you know.

                    HIPPO
For what? I've just been through hell. You, by the way,
look angelic.

                    REBECCA
So who'd you go out drinking with?

                    HIPPO

John.

                    REBECCA
I don't like him.

                    HIPPO
He's my agent. You should like him.

                    REBECCA
He doesn't seem to be doing much good for you.

                    HIPPO
He says that's my fault.

                    REBECCA
Does he have any leads?

                    HIPPO
No. Nothing. I think I'm jinxed. You know, like a hex?

                    REBECCA
Were there women around?

                    HIPPO
Yeah. Can you believe they let 'em in bars?
          (REBECCA turns away.)
But I didn't even look at 'em, because you, Becky, are
my one true love.
          (She turns back to face him.)
I'm here, aren't I?

                    REBECCA
Yeah, five hours late. Why didn't you call?

                    HIPPO
I don't know. I wanted to see if John had something for
me and . . . I don't know. I should've. I'm sorry.

                    REBECCA
You know you have the most ridiculous-looking bald
spot.
          (HIPPO looks in the mirror. REBECCA motions
          to the bat on the floor.)
You better throw that in the garbage. I don't know that
I'm not still mad at you. The strangest things happen
to you.

                    HIPPO
I was serious about what I said. I'm gonna read up on
hexes. Meanwhile, can I try to seduce you?

                    REBECCA
You can try--if you wash your head.

                    HIPPO
Aren't I sexy?

                         REBECCA
At the moment you look a little unbalanced.

                          HIPPO
Well that's sexy isn't it?

                         REBECCA
Go throw the bat away. I'll be in the bedroom.
            (She starts toward bedroom.)

                          HIPPO
            (He starts to follow her.)
Can't I just throw it away tomorrow?

                         REBECCA
Now.

                          HIPPO
Sorry about messing up.

                         REBECCA
Yeah, I know. John'll get something for you soon.

            (REBECCA GOES INTO THE BEDROOM. HIPPO picks
            up the bat, runs to the kitchen, runs out of
            the kitchen without the bat, and flicks off
            the living room lights as he TURNS INTO THE
            BEDROOM.)

                        THE END

EVALUATION

If I were to ask "What was that scene about?" most people would probably respond, "It's about a man with a bat caught in his hair." They would be right. That unusual and striking incident gives focus to the scene. Interestingly, however, the author chose *not* to show us the attack of the bat, but its aftermath. The attack of the bat might be visually exciting, but the interplay of the characters is of more lasting interest. What is most impressive to me is that the student has not simply dramatized an incident. Rather he has taken the incident, added to it specific characters, and developed from it a situation of some complexity. The result is a fascinating and cohesive scene.

The scene gets off to a good start with a strong attack that lets us know immediately that something is wrong. It is no accident that the first five sentences are questions, and those questions set up conflict between the characters and provide interest for the audience.

After Rebecca's "Dammit, Hippo, where have you been?" we know the central conflict of the scene. Her "Drunk again, aren't you?" strengthens the conflict and begins to tell us something about Hippo's character. Because of the unusual name Hippo, we are also cued to the fact that there may be some unusual elements to this scene. Rebecca's third question, "What'd you do to your head?" which is asked and then repeated after Hippo's first line, serves two functions. It focuses attention on what will be a central element of the scene (the bat), and it shows right away that Rebecca is concerned for Hippo even when she's mad at him.

Hippo's "Where's the scissors?" lets us know there's something in his hair, and we're prepared for something sticky, like gum, that needs to be cut out rather than washed or brushed out. Finally Rebecca's declaration, "You've got a bat in your hair!" does what good drama should do; it both confirms and surprises. It confirms that something is wrong with Hippo's head, and it surprises us with what that something is.

The author has not forgotten about action, conflict, character, and dialogue. He has the physical action of finding the scissors and of cutting out the intruder, as well as the less physical but more important action of settling the conflict between Rebecca and Hippo. That conflict is nicely handled. At first Rebecca is obviously angry. Then she is distracted, first by the bat and then by the search for the scissors and the mini-argument over who put them away. She returns from the scissors distraction to the bat distraction and demonstrates a genuine interest in what happened. Her interest turns to concern at the possibility of Hippo's being hurt by the car or by the bat. Having dealt with the extended distraction of the bat, Rebecca returns to her anger. But by this time it is more than somewhat dissipated; only her jealousy about the presence of other women allows her to keep any heat at all. That too, finally, is dropped after Hippo's apology.

The conflict is resolved as Rebecca forgives Hippo. That resolution succeeds because the author manages to develop both the relationship and the situation. First, these two people care about each other. Hippo knows he has erred, and his apology is sincere. Rebecca's concern for Hippo emerges even in the midst of her anger.

Nevertheless, Rebecca might not accept Hippo's apology so speedily without the slight but careful delineation of the situation. Life has not been going well for Hippo. He seems to be living under a dark cloud, and the attack of the bat fits the pattern of misfortune that has been following him. Rebecca is sensitive enough to realize that Hippo is going through a difficult period. As a result, she accommodates him perhaps a bit more readily than she might otherwise.

The characters are differentiated nicely in terms of dialogue. Rebecca is direct in her lines and in the manner in which she approaches problems: "Were there women around?" "It doesn't look like it bit you. I don't see any blood." Hippo is much less direct. Several times he simply does not respond to direct questions. When he does, he uses tactics to diminish the significance of the answer. Hence, in response to "What'd you do to your head?" he says, "Nothing. Where are the scissors?" In response to "Were there women there?" he jokes, "Yeah. Can you believe they let 'em in bars?"

What problems are there in this scene? Only minor ones. The author could do more with the finding of the scissors. Perhaps they can't find the scissors and he has to use a knife. Why do that? Whenever possible, complicate the lives of your characters. Always make things harder for them, never make things easier.

A final reason I like the scene is that it makes us want to find out more. What is Hippo going to do about his jinx? Frequently one of the strengths of a good scene is that it will lead you in other directions.

Now let's look at a scene that sprang from a student's intimate knowledge of an environment and her careful observation of character.

EXAMPLE

THE DONUT SHOP
by Andrea Fisher

THE SETTING is a donut shop. There is an L-shaped counter center. Stage right are three one-piece plastic table and chair sets. Behind the counter are various machines and dispensers. The donuts are on display in an alcove in the upstage wall. Small swivel stools are spaced regularly around the counter. There is a cash register on the upstage end of the counter. Behind the counter stage left is a doorway to the kitchen. The street entrance and exit is downstage right. The restaurant has the atmosphere of a truck stop, except that it is brighter and more plastic than most. The main colors are red and white. The restaurant light is bright and fluorescent. The time is seven P.M. on a Friday night in early August.

(Darkness. BILL's voice comes out of the darkness and the lights fade up as he speaks.)

BILL

Well, I'm walking behind her, y'know, kind of bird-dogging her. About this time, she starts slowin' down. And then I start slowin' down too, 'cause you know, this is a big woman, Harlan, and if she doesn't like somethin' you're doin', she doesn't hesitate to let you know. Yolanda may have been a lot of things, but she wasn't shy. So anyway, here we are, her slowin' down, and me slowin' down, and I'm wonderin' if we're both gonna come to a dead standstill or what's gonna happen, when suddenly she turns around and looks at me kinda

lazy and says "Warm, ain't it?" Can you beat that? Here
I am, about ready to melt into a little puddle right
there in front of her, and she says, turnin' around
real slow, you know, with her eyes half-closed, "Warm,
ain't it?" Whew, I'm tellin' you!

>   (The lights reveal BILL and HARLAN at a
>   table. BILL is a big man who tries to spread
>   good cheer. He always has a story and
>   jovially repeats himself if he has nothing
>   else to say. HARLAN is short and muscular,
>   with bright eyes. He is always filthy--he
>   works at a garage--and he wears a cap with
>   an oil company insignia. His speech is often
>   unintelligible. Much of his conversation
>   consists of a series of insane cackles and
>   whoops, which usually result in his choking
>   on his own phlegm. He is a good-natured
>   animal, and his wild laughter is
>   particularly provoked by anything remotely
>   connected with sex.

>   ENTER EMMETT. EMMETT has the unmistakable
>   look of a misfit. He carries himself with an
>   open, slightly lunatic air of goodwill. He
>   has a sense of drama, and is given to
>   mysterious pauses and knowing nods. He
>   speaks loudly and smiles a lot. He is 30
>   years old, but seems younger. He carries a
>   small bag in one hand.)

HARLAN

Hey, Emmett, when you gettin' married?
>   (EMMETT grins, sits at the counter, takes a
>   napkin from a holder and spreads it
>   carefully in front of him.)

BILL

Yeah, Emmett, when's the happy day?

EMMETT

>   (Catching sight of a fly)

Ah, there's that little booger now.
>   (He stalks the fly, and after a series of
>   unsuccessful thrusts with his hand, finally
>   succeeds in catching it.)

Aha!
>   (He takes it to the door and lets it go.)

                              BILL
You're really deadly there, Emmett.

                             HARLAN
Is that how you gonna catch you a wife?

                             EMMETT
             (Singing in a pleasant <u>sotto voce</u> as he
             fiddles with the salt and pepper)
Gimme that old time religion,
Gimme that old time religion,
Gimme that old time religion,
It's good enough for me.
             (Calling)
Hey, how do you get some service around here?

                             HARLAN
Yeah, Emmett, we know who you want service from.
             (He explodes into laughter as CEECEE ENTERS
             from the kitchen. She wears a white waitress
             uniform and a red apron. She is 16, very
             pretty, and a tireless and heavy-handed
             flirt.)

                             CEECEE
You guys talkin' 'bout me?

                              BILL
Well, maybe Ceecee.

                             CEECEE
Now Bill, you just watch what you're saying about me.

                             HARLAN
Ceecee, we was just talkin' 'bout what nice legs you
got.

                             CEECEE
Well thank you, Harlan. Hiya, Emmett. Haven't seen you
for at least a coupla hours. Where ya been?

                             EMMETT
Oh, around.

                             CEECEE
Whatcha got in the bag, Emmett?

                             EMMETT
Oh . . . things.

                              BILL
What do you think of Emmett, Ceecee?

                    CEECEE
Oh, I think he's really somethin'.

                    BILL
We been tellin' him he'd make some girl a fine husband.

                    CEECEE
I bet he would, too. Who you gonna marry, Emmett?

                    EMMETT
          (Embarrassed)
I ain't gettin' married.

                    BILL
Aw, come on, Emmett. A grown boy like you ought to
settle down. Now when's it gonna be?

                    EMMETT
I tell ya, I ain't never been married, I ain't married
now, and I ain't never gonna be married, and that's all
there is to it.
          (As punctuation he yanks from his bag a doll
          that vaguely resembles CEECEE and sets it on
          the counter.)

                    BILL
Where'd you get the doll, Emmett?

                    EMMETT
Won it at the fair.

                    CEECEE
Oh, Emmett, isn't that cute. You brought me a doll!
How'd you know it was my birthday?

                    EMMETT
It ain't your birthday. It's my birthday.

                    CEECEE
Well mine's next week. When's yours?

                    EMMETT
Today!

                    CEECEE
Today, huh? Well I just think it's awful mean not to
give me that doll. He won't give it to me, Bill, and
I've always wanted one of those dolls.
          (She grabs the doll.)
Why look. It looks just like me. I could be its mama.

                    EMMETT
           (Grabbing it back)
I don't see how.
           (HARLAN starts to laugh.)

                    CEECEE
Oh shut up, Harlan.
           (HARLAN attempts to stifle himself as CEECEE
           pouts.)
Who you gonna give it to, Emmett?

                    EMMETT
I'm not giving it to nobody.

                    CEECEE
Well, what good is a doll to you?

                    EMMETT
You never know. You never know.

                    CEECEE
You're just downright mean, Emmett.

                    EMMETT
I'm not mean. I just . . . well . . . what . . . what
are you giving me for my birthday? Nothing I'll bet.

                    CEECEE
Oh, I'll think of something.
     (HARLAN whoops and CEECEE snaps a towel at him.)
Hush, Harlan.

                    BILL
Well, while you're thinking, Ceecee, could you get me a
cup of coffee?

                    CEECEE
Okay. You want anything, Harlan?

                    HARLAN
I sure do!
           (He whoops and starts laughing again.)

                    CEECEE
Oh, shut up! I mean you want anything to drink?

                    HARLAN
Yeah, get me a Pepsi, I guess.
           (CEECEE gets the drinks as the scene
           continues.)

                    BILL
Well, Emmett, when're you gonna stop playin' with dolls
and get you some of the real thing?

                    EMMETT
Aw, come on, Bill.

                    CEECEE
Well, that's it! I know what I'll give Emmett for his
birthday! Emmett, I'm gonna give you one kiss for every
year old you are, just as soon as I get off.

                    HARLAN
Hoo-wie! You heard it, Emmett. You gotta marry her now!

                    EMMETT
          (Overlapping Harlan)
Oh no you won't, Ceecee. You're not even gonna get near
me. No sir!

                    BILL
Well, I call that one DEE-luxe birthday present,
Emmett.

                    EMMETT
          (Rocking back and forth and giggling)
Oh no. Oh-h no she won't.

                    CEECEE
Yes I will too, Emmett, just as soon as I get off, so
you better pucker up.
          (BILL and HARLAN laugh as EMMETT continues
          to rock and protest.)
I just don't know how I'm gonna control myself till
then. I just don't know how. . . .

                    BILL
Why don't you just do it now, Ceecee?

                    CEECEE
Huh?

                    BILL
The boss ain't around. And I don't think anybody here's
gonna object if you plant one on Emmett.

                    CEECEE
Well, I don't know if I should . . . you know . . .
till I get off.

                    HARLAN
Come on, Ceecee. Give the boy his present.

CEECEE
(Gets a comb and compact from her purse
behind the counter)
Well let me make sure I look all right.

BILL
Hurry up, Ceecee, the poor boy's chafing at the bit.

CEECEE
Maybe I oughta go put on some rouge.

BILL
Whatsa matter, girl? Emmett too hot for you to handle?

HARLAN
(Going behind EMMETT and putting his hands
on his shoulders)
He's all yours, girl. Come and get him!

EMMETT
Oh-h, no!

CEECEE
Well. . . . What d'ya think, Bill, should I take my
apron off or leave it on?
(HARLAN hoots.)

BILL
Well, I think you oughta at least loosen it a little.
You know, wear it kinda low.

CEECEE
Huh?

HARLAN
Come on, Ceecee, it's now or never.

CEECEE
Okay. Okay, Emmett, ready or not, here I come.

EMMETT
Oh-h, no!
(He springs up.)

CEECEE
Oh, yes, Emmett. Here I come.
(CEECEE comes slowly around the counter
saying "Here I come. Get ready, Emmett."
Finally EMMETT lurches forward and grabs his
doll.)

                    EMMETT
Oh-h, no! Oh-h, no!
          (He dives over the counter and RUNS OUT THE
          DOOR.)

                    HARLAN
That was some birthday present, Ceecee.

                    BILL
Well what d'ya know? Just about every man between here
and Shanghai trying to get it, and Emmett turns it
down.
          (HARLAN cackles, and CEECEE flings a dish
          towel at him.)

                    THE END

EVALUATION

A major difference between this scene and the previous one is that "The Donut Shop" has four characters and "Hippo and the Bat" has only two. As you know from previous work, relationships are either direct or modulated. Rebecca and Hippo have a direct conversation. What Rebecca says is meant for Hippo, and what he says is meant for her. Even when a line such as "Were there women around?" has underlying meaning, both the surface and the underlying meanings are exchanged between just those two characters.

As the number of characters increases, the structural possibilities increase geometrically, as you saw in Chapter 5. Bill has a relationship with Harlan, and Harlan has one with Bill. Each of those relationships is altered by Emmett's entrance. In addition, Emmett has a relationship with each man, and each man with him.

All of that becomes even more complex with Ceecee's appearance. She has a relationship with each of the three other characters, and each of them has a relationship with her. All of those relationships are affected by the presence of others in the room. Without Bill and Harlan around, for example, Ceecee might tease Emmett a bit or she might ignore him, but I doubt she would approach him as if to kiss him. She does that because Bill, by calling her bluff *in front of others* dares her to it, and she cannot back down without losing credit. Her performance is at least as much for Bill and Harlan's benefit as for Emmett's.

This scene lives by its characters. The author has observed carefully and has selected telling details about each of them. There's Emmett, a grown man, but childlike in his simplicity. Like an adolescent, he is fascinated by Ceecee and yet scared of her. His very common language such as "Oh . . . things," and "You never know" has a mysterious imprecision about it.

Ceecee is at that stage where a blossoming young girl tests her effects on men. She's flirtatious. She wants people to think she knows everything about love and life. She wants the attention of men, but she also wants to control the relationship.

Bill is a relatively normal individual who provides links between the characters. He appears to have a good instinct for human nature, and he occasionally turns a clever phrase.

Harlan seems sharper mentally than Emmett, but emotionally he is hardly more mature. His cackles and whoops sharply define his character. Collectively Bill and Harlan provide an audience before whom Ceecee and Emmett can act out their drama.

The author could have written this scene with three characters, omitting either Bill or Harlan, but I think both are necessary. Harlan gives us the crudest and most basic possible vision of relationships—everything is based on sex. Bill has a more sophisticated sense of human relationships, which is why he is the one who eventually calls Ceecee's bluff.

But characters, by themselves, do not make a scene. They must be spun into action, into conflict. In that respect the writer has developed these individuals carefully and cleverly. Emmett's singing and fly-catching instantly mark him as an eccentric. As in some other scenes, questions immediately provide an opening hook. Although they are asked in jest, the questions put to Emmett about marriage and a wife—and his ignoring those questions—establish character interactions and focus the scene on male-female relationships.

The author does not try to introduce all the characters at once. We first meet Bill and Harlan. Then Emmett. Then Ceecee. Such a progression allows the audience to get to know the characters more easily than if all four were introduced together in a group. Virtually as soon as all the characters are identified and Emmett has answered the opening question ("I ain't never gonna be married!"), the doll is brought out. Like the sweater, the cookies, and the bat in previous scenes, the doll inaugurates the conflict and becomes a concrete symbol of it. Ceecee pursues it not only to tease Emmett but also to prove to herself and to Bill and Harlan that she controls the relationship.

When Emmett asks her for a present, Ceecee comes up with the kiss idea, knowing it will both tantalize and embarrass Emmett. She also thinks she will never have to follow through on it.

The key moment of the scene occurs when Bill says, "Why don't you just do it?" Ceecee is caught. She backpedals and delays, but she finally goes ahead. She doesn't want to lose face with Bill and Harlan. She also sees that the thought of a kiss from her is unsettling Emmett, and Ceecee likes to unsettle men. The playwright has remembered the first lesson; she has put the characters into action. The picture of Ceecee torturing Emmett, slowly approaching with her lips puckered as Bill and Harlan attempt to hold him, has solid theatrical value.

Having built to the climactic moment of the kiss, the author deflates the bubble by having Emmett bolt. In a way, all the characters in this scene win. Ceecee confirms her power over men and retains her standing with Bill and Harlan. The two men achieve a good laugh and teach a small lesson to Ceecee in the process. Emmett avoids contact with a girl, but at the same time he has the joy of knowing that Ceecee was really going to kiss him.

No final resolution is achieved. The scene leaves us wondering whether Emmett will ever get his kiss or if he'll somehow get revenge on Ceecee for taunting him. If a scene leaves you wanting to know more, as I think both "The Donut Shop" and "Hippo and the Bat" do, that's a good sign. And that's when scenes begin to grow and expand into plays.

7

• • • • • • • •

Writing from a Source

EXERCISE 7 • WRITING FROM A SOURCE
Write a scene using an item from a newspaper or a magazine as
the springboard. Use as many characters and as much dialogue
as you need.

In this chapter your assignment is to write a scene using an item
from a newspaper or magazine as a starting place. A historical incident
would also work. It may seem that we're heading in the wrong direction.
After all, in Chapter 6 you learned that everything you write must come
from your own experience, from your own thoughts and feelings.

That's still true. But there are different ways to get at your
thoughts and feelings, and the purpose of this chapter is to work in the
opposite direction from the preceding exercise. In the exercise in Chapter
6, the impetus is internal. The germinal idea is something you're very familiar with, and, in the course of developing that idea into a dramatic
piece, you must provide it an external reality that other people can
recognize.

In this exercise the seed is something *external*—a news item, for
example—and you must make it *internal*. You must create environments,
situations, conflicts, and characters with depth and reality. Perhaps giving
"reality" to a news item or a historical incident that begins with "reality"
sounds redundant. Or at least easy. Yet you will be amazed how easy it
is to begin with a "real" incident and wind up with an "unreal" play.

Authors writing about George Washington or John Merrick (the
so-called "Elephant Man") cannot merely relate incidents. They must
create from their own experiences—which in these cases includes readings about Washington or Merrick—a Washington or a Merrick that is an
understandable human being to an audience. That, of course, is what
Maxwell Anderson tried to do in *Valley Forge* and what Bernard Pomer-
ance tried to do in *The Elephant Man*.

In one of my classes a student tried to write a scene based on an
Arab terrorist. The student attempted to develop the human side of the
young activist. The scene broke down because the young writer was unable to breathe "reality" into this "real" person. The writer had little sense
of religious fervor, of political calculation, or of group honor, any of
which might propel the terrorist to action. Some research, some knowledge of Islam, of Mideast politics, or of the psychology of terror might

have helped the writer provide a specific and dramatically viable focus to the scene.

Another student wrote a scene about a wealthy businessman who had committed a crime. But his dialogue about "big deals" and "lots of money" was so general that the businessman immediately lost all credibility and seemed fake.

In other words, if you write a scene about a terrorist or a businessman or even George Washington, you must come to know that person like a good friend or a next-door neighbor so that you can show us how that person thinks, talks, and acts.

Examples of plays that proceed from real events to dramatized events are legion. Peter Shaffer used historical material about Pizarro and the Incas as a basis for *The Royal Hunt of the Sun.* Later Shaffer was intrigued by a news account he heard about a boy who put out the eyes of a stable of horses. Shaffer wondered what could have possessed him to do that. The playwright invented a psychiatrist to probe the young boy's psyche, and eventually the doctor became as important as or perhaps more important to the play than the boy. Shaffer's imagination created the circumstances and his dramatic skill created *Equus.* Shaffer returned to historical material in *Amadeus,* using as his seed the old musician's rumor that Wolfgang Amadeus Mozart had been murdered. Just as the psychiatrist in *Equus* emerged as a central figure, so in *Amadeus,* as the play developed in Shaffer's mind, the apparently subsidiary character of Salieri became the linchpin of the piece.

Lee Blessing used an account of nuclear disarmament discussions as a seed for his play *A Walk in the Woods,* and he latched onto anecdotes about Ty Cobb's apparently racist opinions in writing *Cobb.*

In *The Crucible* Arthur Miller used news items and historical material in two ways. Enraged by the Communist-hunting tactics of the House of Representatives Committee on Un-American Activities in the late 1940s and early 1950s, he wrote *The Crucible.* The play dealt with the same topics that Miller saw treated every day in the newspaper—persecution, guilt by association, and a kind of mass hysteria. But instead of setting his play in the 1950s in Washington, D.C., he set it in 1692 in Salem, Massachusetts, during the days of the Puritan witch hunts.

Just as there is a danger, when working with material that comes from your own experience, in hewing too much to actual fact, so there is a danger in working with external materials in trying to be *too* factual. You *can* change history. Shakespeare did it all the time. There are, however, some factors you might consider. Generally, the closer in time and the more prominent the incident, the less leeway an author has. If you want to write about the explosion of the *Challenger,* you could hardly use any names other than those of the actual astronauts, and the incidents you incorporate would need to parallel actual events pretty closely. Even at that, you would certainly be inventing dialogue and action based on your understanding of the people and events.

If you were writing about the competition sparked by the beginning of human flight, you'd have to use Orville and Wilbur Wright, but the various people around them are virtually unknown now. The distance in time allows greater dramatic license. You could probably invent or combine characters or even create or alter incidents, depending on your dramatic purpose.

I have a final warning regarding historical materials. Frequently when dealing with historical personages, writers lose their sense of language and dialogue. Suddenly everyone is expressing grand thoughts in complete sentences with no contractions. Yes, dialogue must reflect the time period of a play, but it must first give us entry to an apparently real human being. My advice is to write the scene as if it is being spoken today, and then eliminate obvious modern language in revisions.

Often when I assign this exercise a student will ask me about doing an adaptation of some other work, such as a scene from a novel or a short story. Adapting a story from one medium to another makes a valid and interesting exercise, but it is a very different exercise from this one. When you adapt something, the story itself is complete. The challenge is to move that story from one medium to another. It's as if you had a complete jigsaw puzzle, and you were trying to redo the picture as an oil painting or a watercolor. In this exercise the story is *not* complete. You must create at least part of it. Think of it as having one piece of a jigsaw puzzle, and you are being asked to design a picture around that one piece.

At the start of this chapter, I said that working from a source amounted to taking something external and making it internal. In fact it's really a three-part process. First, you take something external, such as a news article. Second, you make it internal by imbuing it with knowledge, with thought, and with feeling. Third, you translate it into a scene or a play, thereby making it once again into an external product.

Now let's observe that process at work in two student-written pieces.

● ● ● **EXERCISE 7**
Write a scene using an item from a newspaper or magazine as the springboard. Use as much dialogue and as many characters as necessary. Your scene should be at least 5 pages long. For comparison purposes, be sure to keep the news item along with the scene.

The scene should not attempt simply to reconstruct what happened. Rather, allow your imagination to play with the account until you find something dramatically interesting to work with.

EXAMPLE

NEWS ITEM: "A Face From the Past" (from <u>Newsweek</u>)

Rudolf Hess was alone on his birthday last month--as he is almost every day of the year. At 83, Hitler's former deputy is the sole remaining

inmate of Berlin's Spandau Prison and, under an
agreement among the World War II victors, he is
spending a lifetime in solitary confinement. Except for a monthly chat with his son, Wolf, Hess
is permitted no visitors, and when he became a
grandfather for the first time not long ago, he
got the word by telegram. No one is allowed to
photograph Hess, but despite the stringent secu-
rity, he recently managed to pose for a candid
shot--presumably taken by a guard. Last week the
photo became available.

Age and his 31 years in Spandau have left an
unmistakable stamp on Hess. Members of the prison
staff describe him as "a broken old man," and
last February he attempted to kill himself with a
knife. British, French, and American officials
say they would be willing to consider a reprieve
for Hess, but the Soviet Union refuses. Notes the
Soviet journal Literaturnaya Gezeta: "We remember
everything about the past."*

A FLASH OF LIFE
by Brian Ruberry

THE SETTING is a prison cell lit by the
early morning sun shining through a lone
window. In the cell sits a frail-looking OLD
MAN on the edge of his cot. He has just
awakened. Across from him is a rusty sink,
and an empty metal frame for a mirror is on
the wall above the sink. An open toilet is
in the far corner and it can be heard
running. The man's full head of hair is snow
white, and he wears a gray prison uniform.
The OLD MAN slowly, painfully arises and
jiggles the toilet handle. After no success,
he sits back down and puts his face in his
hands.

A short, plump PRISON GUARD ENTERS quietly
with a camera in one hand, and his words
startle the man.

GUARD
(Putting his head against the bars)
This ain't no time to be poutin', Mr. Hess. You already
forgotten what day it is?

(The OLD MAN slowly shakes his head.)
Oh, come on now! What happened a year ago today, huh?

OLD MAN
(Shaking his head and speaking slowly)
I don't know.

GUARD
Don't, huh? Well it's your birthday, Mr. Hess. Your
eighty-third birthday. Now how 'bout that?

OLD MAN
(Looking up)
Today? My birthday?

GUARD
That's right. Now I ain't got no present or nothing,
but I did compose you a little birthday song. Wanna
hear it?
(The OLD MAN nods.)
Huh, Mr. Hess?

OLD MAN
Yes. Yes.

GUARD
All right.
(He takes a folded paper from his shirt
pocket.)
But remember, I ain't no songwriter or nothing. So
don't expect much.
(He clears his throat loudly and sings.)
    Happy birthday to you.
    Happy birthday to you.
    You killed all them Jews.
    Now they're killin' you.

You like that?
(The OLD MAN nods his head and the guard
laughs as he opens the cell door and
enters.)
Since you like my song so much I got one more treat for
you before you take your walk.
(Holding up a camera)
You know what this is?

OLD MAN
A camera.

GUARD
That's right! Now I'll tell ya what. If you keep this
quiet, I'll snap a picture of the birthday boy.

                    OLD MAN
It's forbidden. It's against the rules.

                    GUARD
              (Waving his hands)
It ain't against the rules if it's the warden who's
ordering the picture to be taken. Ya see, I'm just
carrying out his instructions.

                    OLD MAN
I don't want a picture.

                    GUARD
Now don't be bashful. No one's gonna recognize you
anyway.

                    OLD MAN
I don't think I look so good anymore. I'd be ashamed.

                    GUARD
              (Bending over to him in a confidential
              manner)
Haven't I been your friend all these years? Haven't I?
Now would you want your friend to get in trouble for
not carrying out the warden's orders? Would you?
              (The OLD MAN shakes his head.)
Well all right then. Just sit up straight and it'll be
over in a flash.

                    OLD MAN
              (Running his fingers through his hair)
My hair. My hair is not so neat.

                    GUARD
              (Giving him a comb from his pocket)
Here. Now hurry up 'cause you're cutting into your
walking time.

                    OLD MAN
              (Rises and faces the empty mirror frame)
It's broken. My mirror's broken. They never replaced
it.

                    GUARD
              (Laughing)
Never replaced it? Hell, that was twenty years ago!
Here, it don't matter. I'll do it.
              (He strokes the comb through the OLD MAN's
              hair a few times.)
There. If I didn't know better, I'd swear I was
standing next to Elvis Presley hisself.

                          OLD MAN
                (Looking down at his clothes)
My clothes.

                          GUARD
             (Impatiently straightens his clothes)
All right, all right. Your clothes are fine. Now sit
down.
                (He indicates the cot, and the OLD MAN is
                impassive as he crosses his arms over his
                chest and the camera flashes.)
There. Now that didn't hurt, did it?
                (He checks his watch.)
Hurry up, 'cause we ain't got much time left. The way
you walk I might die of old age my damn self.
                (The GUARD hands the OLD MAN a cane and they
                begin to leave the cell.)

                          OLD MAN
Do you think we could walk the eastern path today? We
haven't taken that path in a long time.

                          GUARD
Eastern path, huh? I bet you would like to take it
considerin' we done sold it back to the state. You know
that, Mr. Hess. We sold it back even before I came
here. Hurry up now. You walk like a dead man. Come on.
Time's a wastin'. Time's a wastin'.
                (THEY EXIT.)

                        THE END

EVALUATION

The article and the scene both relate the account of a guard taking
a picture of an old Nazi. Notice that the news item tells us *about* the man.
The scene shows us the man himself! Or at least one author's version of
that man.

As with much drama, the most striking elements of the scene are
the most human elements of the scene. At first we see an apparently confused old man imprisoned in a bleak cell. The guard taunts him and
forces him to sit for the photo. Amazingly, perhaps, the author generates
sympathy for the Nazi!

But then other qualities come out. The old man's vanity shows
through in his concern for his hair and his clothes, and the audience
might be reminded of a dapper young man in a sleek uniform.

Finally, without hitting us over the head, the author carefully
makes his point. "I'm just carrying out his instructions," the guard says
of the warden. "Would you want your friend to get in trouble for not
carrying out the warden's orders?" Of course, the guard's question

echoes the defense of many Nazi war criminals at the Nuremberg trials, and it does not excuse the young guard any more than it excused the German officers.

EXAMPLE

NEWS ITEM: The student used an advertisement from the kind of tabloid newspaper sold at the check-out counters at supermarkets. This particular item advertised a special pendant that promised to bring good luck.

BAD LUCK
by Charles Hannon

THE SETTING is a park. A WELL-DRESSED MAN is sitting on a park bench reading the <u>Wall Street Journal.</u> ANOTHER MAN ENTERS. He is disheveled. A sport coat dangles from one hand, his tie is loosened, and his hair is mussed. As he walks near the bench, he stops. He has stepped in something.

DISHEVELED MAN

Shoot!

(The man on the bench looks up. The DISHEVELED MAN picks up his right foot. He pushes the sole against the ground and wriggles it back and forth. He slowly picks it up again.)

Darn!

(He hops over to the bench and sits. He inspects the sole of his shoe.)

Oh, man.

(He looks on the ground for a stick, finds one, and begins to scrape at his shoe. He speaks to the rhythm of his scraping.)

This . . . is the worst day . . . of my life.

WELL-DRESSED MAN

Bad one, huh?

DISHEVELED MAN

First my car goes dead. In the middle of the freeway. Then I go to work, and my boss fires me. So I take a little walk in the park to relax, and I get robbed. And now this! I mean this is the final straw.

WELL-DRESSED MAN

I feel for you. I used to go through days just like that.

                    DISHEVELED MAN
You did?

                    WELL-DRESSED MAN
All the time. Lost my job. Had my car stolen. Listen,
if there was one bird flying in the sky, you know where
he'd drop his load? Right on me!

                    DISHEVELED MAN
That's just how I feel.

                    WELL-DRESSED MAN
A car drives by and hits a puddle. I'd be the one
that'd get splashed.

                    DISHEVELED MAN
Exactly!

                    WELL-DRESSED MAN
You know what your problem is?

                    DISHEVELED MAN
No. What?

                    WELL-DRESSED MAN
Somebody probably put a hex on you.

                    DISHEVELED MAN
What?

                    WELL-DRESSED MAN
A hex. You know what a hex is?

                    DISHEVELED MAN
Yeah. I know what a hex is.

                    WELL-DRESSED MAN
                  (Overlapping)
So bad things would happen to you.

                    DISHEVELED MAN
Are you kidding?

                    WELL-DRESSED MAN
Not at all.

                    DISHEVELED MAN
You don't really believe in hexes, do you?

                    WELL-DRESSED MAN
I didn't up until a few months ago, but then my whole
life turned around.

                    DISHEVELED MAN
It did?

                    WELL-DRESSED MAN
Well, like I said, I'd lost my job. Well I got a job
that pays twice as much. I mean, I invest in a stock
and . . . bing! It goes through the roof! I know it's
hard to believe.

                    DISHEVELED MAN
It is a little.

                    WELL-DRESSED MAN
You married?

                    DISHEVELED MAN
Well . . . ah . . . yeah. But last week my wife said
she's seeing another man.

                    WELL-DRESSED MAN
Six months ago my wife left me. I was miserable. But
now there's three very wealthy, very attractive women
beating down my door. You don't look like you believe
me.

                    DISHEVELED MAN
It's just that a hex is a little . . . (trails off)

                    WELL-DRESSED MAN
Weird? Well, look at this way. Don't you have any
enemies? Isn't there anyone you know who would benefit
from your misfortunes? Could there be someone out there
who would just love to see the world dump on you?

                    DISHEVELED MAN
Well . . . maybe.

                    WELL-DRESSED MAN
Sure there is. You see what I mean?

                    DISHEVELED MAN
But a hex?

                    WELL-DRESSED MAN
What explanation do you have?
                (Laughing lightly)
You really think it's just a coincidence that all these
things happen to you?

                    DISHEVELED MAN
Well . . . I don't know.

                    WELL-DRESSED MAN
What's your birthday?

                    DISHEVELED MAN
April third.

                    WELL-DRESSED MAN
Oh, an Aries.
            (He picks up a local daily paper from the
            bench beside him and looks at it.)
Well, look at that.

                    DISHEVELED MAN
At what?

                    WELL-DRESSED MAN
Your horoscope. "Opportunity presents itself to
overcome problems. Take advantage of it."

                    DISHEVELED MAN
            (He takes the paper and looks intently.)
That's incredible.

                    WELL-DRESSED MAN
Of course, I don't believe in those things myself.

                    DISHEVELED MAN
Neither do I.

                    WELL-DRESSED MAN
Look at Leo. That's mine.

                    DISHEVELED MAN
"Be especially careful today. It's not a time to start
new ventures."

                    WELL-DRESSED MAN
I've had a great day.
            (He pulls a lottery ticket out of his
            pocket.)
Look at this. I even won 50 bucks on a lottery ticket.
They're wrong as often as they're right.

                    DISHEVELED MAN
            (He looks at the newspaper, then at the
            WELL-DRESSED MAN.)
So you thought you were hexed.

                    WELL-DRESSED MAN
I was hexed.

                    DISHEVELED MAN
Okay, you were. Let's suppose for a minute . . . that
I . . .
            (He struggles to say this.)
. . . that I . . . ah . . . actually am . . .

                    WELL-DRESSED MAN
Hexed?

                         DISHEVELED MAN
Yeah. What did you do about it?

                        WELL-DRESSED MAN
I can't tell you.

                         DISHEVELED MAN
What do you mean you can't tell me?

                        WELL-DRESSED MAN
I can't tell you. I can't talk about it.

                         DISHEVELED MAN
Why not?

                        WELL-DRESSED MAN
I just can't.

                         DISHEVELED MAN
You go to all this trouble to convince me that I'm
hexed and then you can't tell me what I'm supposed to
do about it?

                        WELL-DRESSED MAN
Right.

                         DISHEVELED MAN
But you already <u>have</u> talked about it.

                        WELL-DRESSED MAN
Only that I was hexed. Not what I <u>did</u> about it.

                         DISHEVELED MAN
Look, I don't know what your game is. . . .

                        WELL-DRESSED MAN
Hey, don't get upset. It's just not allowed. That is
. . . hey, wait a minute.
                    (He reaches in his inside coat pocket and
                    takes out a piece of newspaper, which he
                    carefully unfolds. He looks at it intently.
                    The DISHEVELED MAN tries to look at it, but
                    the WELL-DRESSED MAN guards it and continues
                    reading.)
Wait. I think it's okay.

                         DISHEVELED MAN
What's okay?

                        WELL-DRESSED MAN
Here. Look at this.

                    DISHEVELED MAN
            (He takes the piece of paper and reads.)
"Lose money? Lose a job? Lose a girlfriend? Maybe it
isn't just bad luck." What is this? What's this from?

                    WELL-DRESSED MAN
Some newspaper.

                    DISHEVELED MAN
Are you selling something?

                    WELL-DRESSED MAN
No. Go ahead and read on.

                    DISHEVELED MAN

            (Scanning, he reads bits and pieces.)
"Remove the hex . . . Send $25." Is this what you did?

                    WELL-DRESSED MAN
I can't say anything.

                    DISHEVELED MAN
"We guarantee satisfaction. The hex will be removed and
your luck will change or your money will be cheerfully
refunded."

                    WELL-DRESSED MAN
Look at the last part.

                    DISHEVELED MAN
"This offer is not valid in Utah, Vermont,
Massachusetts. . . ."

                    WELL-DRESSED MAN
The next part down.

                    DISHEVELED MAN
"This offer is invalidated if you tell anyone that you
paid to have the hex removed."
            (He considers that a second, then looks at
            the WELL-DRESSED MAN, who looks sage.)
Is that why you can't say anything?

                    WELL-DRESSED MAN
Draw your own conclusions.

                    DISHEVELED MAN
So you really can't talk about whether you did this or
not, huh?

                    WELL-DRESSED MAN
I can't even talk about whether I can talk about it.

DISHEVELED MAN
(He pauses, then comes to a decision.)
Look, could I borrow this for a few minutes and make a copy?

WELL-DRESSED MAN
Sure. Go ahead.
(The DISHEVELED MAN starts to leave. The
WELL-DRESSED MAN takes a quarter from his
pocket.)
Here. You may need this.

DISHEVELED MAN
Right. Thanks. I'll be right back.

WELL-DRESSED MAN
Fine. I'll be here.
(The DISHEVELED MAN EXITS. The WELL-DRESSED
MAN resumes reading the Journal. In a moment
something falls on his shoulder. He notices.
He looks up. Then, in disgust)
Damn!
(He gets up and starts off after the
DISHEVELED MAN.)
Wait. Hey!
(He trips, then gets up and continues.)
Damn! Hey! Come back!
(He EXITS.)

THE END

EVALUATION

In "Bad Luck" you can practically see the playwright's mind at work. He probably looked at the advertisement and thought, "Why would anyone send in money for this?" Of course, we don't comment on why he presumably spent money to look at this tabloid in the first place. In any case, he took the concepts of good luck and bad luck, and he created a scene in which one man persuades another man that hexes exist, that he's got one, and that he'd better do something about it.

The first part of the play introduces the two men and the problem. The Disheveled Man is essentially in conflict with the world around him, and he finds a sympathetic ear. That part ends with the introduction of the subject of the "hex." In the second section the two men are in conflict as the Well-Dressed Man seeks to persuade the Disheveled Man that he is indeed hexed. That unit ends with the Disheveled Man's acceptance of the concept of a hex and his question "What did you do about it?"

In the third part the men remain in conflict, but the conflict focuses on the Disheveled Man's attempt to get the Well-Dressed Man to tell him what to do. The last, short unit shows us the final switch, when things start to go wrong again for the Well-Dressed Man.

The play has several nice turns and reversals in it, all of which remain true to the characters. The Well-Dressed Man seems normal. He speaks of a hex in the most reasonable manner. That establishes a veneer of normality to what is actually a rather bizarre situation.

When the Well-Dressed Man brings up astrology, the audience might be ready to conclude that he's a certified loon. But just as they begin to write him off, he himself discounts the astrological advice!

Once he has succeeded in getting the Disheveled Man to ask what he needs to do, one rather expects some sort of recruitment or sales pitch: "Buy these self confidence tapes and they'll fix your problem." It's a pleasant surprise when he refuses to divulge any information, and it heightens the apparent mystery of the process, which is appropriate for a scene dealing with hexes.

The resolution of the conflict, with its fractured logic of "I can't tell you what to do, but here's what to do," is fitting. And the final reversal shows us that the author is aware of how difficult it is to disprove folk wisdom. Was the astrological prediction right after all? Was the warning in the hex advertisement accurate? Or is it all just coincidence?

The odd atmosphere that exists in this piece alongside the surface realism continues throughout. Certain lines—"Six months ago my wife left me. I was miserable. But now there's three very wealthy, very attractive women beating down my door" and "Well, look at it this way. Don't you have any enemies? Isn't there anyone you know who would benefit from your misfortunes? Could there be someone out there who would just love to see the world dump on you?"—push the limits of realistic convention. In a completely realistic play, they wouldn't work, but they fit within the style of "Bad Luck."

If there is a weakness to the piece, it is that the characters have little individuality. We learn virtually nothing about these men—not about their jobs, or their cars, or their wives. Not even their names. That they are identified only generically as the "Disheveled Man" or the "Well-Dressed Man" testifies to their lack of specificity.

Having said that, I should also point out that their interaction is very full and complete. Furthermore, added detail might ruin the careful element of unreality by providing them too much surface reality.

You've now written a range of pieces of varying complexity. The next chapter is designed to help you decide where you can go from here and how to get there. You'll be taking your scenes and molding and extending them into plays.

8

· · · · · · · ·

Writing Your Play

EXERCISE 8 • SPONTANEOUS COMPOSITION
Without stopping to edit, write a monologue in which you explore your own thoughts and feelings. Do not be concerned with character, meaning, structure, punctuation, or grammar.

EXERCISE 9 • CHARACTER, PLACE, AND OBJECT
Write a scene using the characters, place, and object that you are given.

EXERCISE 10 • WRITING A PLAY
Write a one-act play about 20 to 40 minutes in length. It can be an extension of a previous scene or an entirely new creation. Use correct playwriting format.

After you've written the scenes required in the preceding chapters, you might be excused for asking, "Okay, do I get to write a play *now?*" The answer is "Yes, of course. In fact, you probably already have!" At the beginning of this book I told you that plays grow like trees. By working through these exercises, you've already begun to nurture your trees.

"But," you protest, "I've only written scenes." What's the difference between a "scene" and a "play"? It isn't the basic elements involved. The elements are the same. A scene and a play both express a story through dramatic means. They both place characters in conflict. They both use dialogue spoken by characters. And, most importantly, they both *animate* the characters. That is, they put them in action. Isn't that what you've done in your scenes?

The difference isn't in the length either. Some very complete plays are very short, and some very long dramatic works are very incomplete. The determining factor lies in that word "complete." More specifically, as I indicated earlier, a scene shows us part of a story. A play shows us the *complete story;* it resolves whatever conflicts have been introduced, and it answers whatever questions have been raised.

If you look at the scenes I've used as illustrations in the preceding chapters, you'll see what I mean. In Chapter 7 "A Flash of Life" could be summarized this way: A guard takes a picture of an old Nazi in prison. That doesn't address the moral of the scene or the atmosphere of the piece, but it does describe the basic action. There is no more to the story. The picture is taken, the conflict is resolved, the characters are gone, the story is finished, and the lesson is conveyed.

Similarly, "Bad Luck" in Chapter 7 could be explained like this: A well-dressed man convinces a disheveled stranger that his bad luck is caused by a hex, and in so doing the man brings bad luck to himself. The piece, of course, is much more than that, but a concise statement helps to focus the significant action of the work.

"Bad Luck," like "A Flash of Life," is a complete short play. It doesn't matter that the Well-Dressed Man goes off to find the Disheveled Man. It doesn't matter that an argument may ensue between them. It doesn't matter whether the Disheveled Man actually sends in the money. The crucial decisions have been made. The turnaround has been effected. Nothing the Well-Dressed Man could say now would alter the situation. The story is complete.

In "Hippo and the Bat," another scene with a hex, in Chapter 6, the immediate problem of the bat is taken care of, and the immediate argument between Hippo and Rebecca is resolved, but there is more to the story. What ultimately will happen with Hippo's streak of bad luck? Will he gain control of his life? Will Hippo and Rebecca sustain their relationship?

"The Donut Shop" in Chapter 6 provokes similar questions. Will Emmett ever get his kiss? Will Ceecee get her comeuppance for her flirtatious behavior? There's more to come in this story.

In Chapter 5, the scene between Beth and her mother represents a complete action: Beth gets her mother's attention. Unless there is some additional complication, that story is over. So is the relationship between Stephanie and Mike in "Nibbles" in the same chapter. Unless the appearance of the unseen Kurt adds some new development, there would be no need for any additional scene.

In "Sis" in Chapter 2, we see the following action: Over her sister's objections a girl takes her sister's sweater to wear on a date. But there seems to be more going on between the two characters. It looks like there's more to the story.

"Good Idea, If It Works" in Chapter 4 presents some interesting questions. As it stands the scene shows us a young tough pressuring his friend into robbing a store. Is that the whole story? Only the playwright can answer that with certainty. But suppose the whole story includes the young tough double-crossing his friend, and the friend, in deep trouble, avenging himself by killing the cause of his problems. Or suppose the story is about the gullible young man and his father. Imagine the father, disappointed in his son, helps him anyway. Through that act the young man learns of his father's love and devotion. The play could be a lesson in the consequences of submitting to peer pressure. It could just as easily be a story of parental support or many other stories. The piece certainly leads us to expect that *something* will happen when the theft is discovered.

In other words, to decide if a work is complete or if it needs additional material, you must first decide *in very concrete terms* just what story you are trying to show.

Let's assume for a second that you've written an intriguing scene, and you've zeroed in on the story you want to dramatize. Your next question might be "What parts of the story should I show?" As we saw in Chapter 2, a story can change radically depending on which parts of it you choose to illustrate. The Greeks understood that. The great tragedians Aeschylus, Sophocles, and Euripides all wrote plays about Electra and Orestes, who revenged themselves by killing their mother, Clytemnestra, and her lover, Aegisthus, because Clytemnestra had murdered their father, Agamemnon, a hero of the Trojan War. Yet the very same characters emerged quite differently in each version of the story because of what the authors chose to show, and those decisions resulted from their understanding of the story.

The Greeks knew something else, too. They knew that where you begin to show a story is extremely important. They also understood that the starting place, or the "point of attack" as it is sometimes called, derives directly from the point of view of the writer—from what the writer wants to emphasize.

Take the story of King Oedipus, for example. Oedipus is born to King Laius of Thebes and his wife Jocasta. An oracle tells Laius that this child will murder his father, so the King disposes of the child on a mountainside. A shepherd finds the child and takes him to Corinth, where he is raised in the household of King Polybus. Upon reaching manhood, he visits the oracle of Apollo at Delphi to divine his future. The oracle warns Oedipus not to return to his home because he will kill his father and marry his mother. Thinking Polybus is his father and Corinth his home, Oedipus runs away to Thebes.

As he approaches Thebes, a man pushes Oedipus off the road, and Oedipus retaliates with a blow that kills the man. Unknown to Oedipus, the man is King Laius, his father. Oedipus continues to Thebes, where he finds the city under a curse. Oedipus saves the city by solving a peculiar riddle, and a grateful city crowns him king. He marries Jocasta. Years later a plague comes upon the city.

All of that is part of the story of Oedipus. But Sophocles does not even *begin* his play until this point, with Thebes under a terrible plague. Oedipus is told that the devastation will continue until the murderer of Laius is found and punished. Oedipus determines to do that, and in the course of the investigation he discovers the horrible truth. In despair, Jocasta commits suicide and Oedipus blinds himself.

Sophocles could have started the play earlier—when Oedipus is born or when Laius decides to get rid of the child or when Oedipus visits the oracle at Delphi. Sophocles could have pursued the story further than he did, after Oedipus leaves Thebes. In fact, Sophocles did just that, but in a separate play, *Oedipus at Colonus,* which shows the story of the end of the old king's life.

Sophocles could have done any of those things, but he realized that where you start the story significantly determines the story you tell. Because Sophocles focused on the end of this story, the main action of the play might be stated as: "Oedipus discovers the truth."

How do you know where to begin your story? Look for the spots where something changes. In *Hamlet* the old king is dead. As bad as that is, a kind of status quo has been established. There is another king on the throne, and the queen has remarried. Shakespeare starts his play at the moment that balance is disrupted, when the ghost of Hamlet's father appears to Hamlet and makes him swear to avenge the murder. Just as Sophocles began *Oedipus* on the day the King decides to find a murderer, so Shakespeare also began his play at the moment a situation changes.

The event that causes the change in the basic situation is referred to in traditional play structure as the "inciting incident." I have refrained from describing typical play structure because I wanted you to explore your own avenues. I didn't want you confined by a rigid system. But now I'd like you to be able to compare your work against traditional play structure. That might give you additional ideas about how to develop your scenes.

A traditional play is said to have a beginning, a middle, and an end. The beginning starts at a point of balance. The characters and the situation are introduced. Even though other events may have occurred earlier, the contending parties are at a pause. In *Macbeth*, a great victory has just been gained. In *The Glass Menagerie* an uneasy truce exists between the family members.

Then a new element is introduced. This is the "inciting incident," and it propels characters to action. It could be a stranger coming into the picture, as in Lyle Kessler's *Orphans*. Or the arrival of the ghost of Hamlet's father, demanding revenge. In *Macbeth,* the witches incite Macbeth's thought of the crown, and then the king arrives at Dunsinane Castle. In *The Glass Menagerie* Amanda discovers that Laura has deceived her about attending typing school.

The inciting incident sets in motion the desires of the main character to accomplish a goal—Macbeth to become king; Hamlet to avenge his father's murder; Amanda to procure a settled life for her daughter. Author Lillian Hellman told aspiring writers that a playwright has about eight minutes to let the audience know who the play is about, what is at stake, where the play is going, and why. That's the beginning.

The desire of the main character meets some form of resistance, and, in the middle section, the play proceeds through a series of conflicts. This is sometimes referred to as "the rising action." Macbeth overcomes his own fears and hesitations. Hamlet tests his uncle's reactions to a play. Amanda pressures Tom to bring home a boy for Laura.

Within the middle part, good playwrights find ways to raise the stakes, to increase the emotional and psychological pressures on the main characters. Not only does Macbeth himself want to be king, but he also receives passionate pressure from his wife. As Hamlet pursues revenge, he is under the additional pressure of a loving relationship with Ophelia, which crumbles. The gentleman caller whom Tom brings to the house turns out to be the one boy Laura has always loved.

As we saw in "The Suit," in Chapter 3, the action often appears to be headed in one direction, but then switches. One side appears to be winning, only to suffer a setback, which is called a "reversal." Just as it appears the girl will get her swimsuit, the mother discovers it costs too much. Macbeth becomes king, but his troubles only multiply. Hamlet stalks his uncle to his chamber, but changes his mind when he finds him praying. Laura kisses Jim O'Connor, but then discovers he's going with another woman.

Finally, at the end, one side wins or loses utterly and decisively and the battle is over. Hamlet kills the king, and he and several others die; Fortinbras is left to pick up the pieces. Macduff slays Macbeth, and Malcolm is crowned the rightful king. Amanda's farfetched dreams for Laura are shattered like the glass figurine, and Tom leaves town.

As you develop your play, always ask yourself: "Why do these characters do what they do *now*?" Most of us procrastinate. We would just as soon put off action, particularly an important, decisive, and difficult action. Characters are the same way.

I will often ask a writer, "Why does this character erupt—or do whatever he or she is doing—just now?" Frequently the answer comes back, "Things built up." I understand that "things build up," but the character didn't erupt the day before. Or the day after. What caused the disturbance on *this* day?

Think about the expression "the straw that broke the camel's back," which means that a camel—or a person—can carry a very heavy load, but eventually a limit will be reached. Physically or psychologically one more item, one more tiny piece of straw will break down the animal.

We, the audience, can't see all the problems that build up, but we want to see that last piece of straw, and you, the author, have to identify it for yourself and for us. In many cases that moment is the beginning of your play.

Whenever possible in the beginning and middle sections of your play, put your characters under pressure to act. If John wants to ask Barbara to marry him, we have a typical situation. If John knows Barbara is also dating someone else, there's more reason for him to speak up. But if John knows Barbara's leaving tomorrow for a job in another part of the country, then John, if he's going to pop the question at all, has to do it *now*.

Why does Macbeth kill his sovereign, Duncan? The prospect of becoming king tantalizes him, to be sure. His wife urges him to it. But beyond the desire and the encouragement, he is presented with *opportunity* and a *time limit*. The king comes to stay at Dunsinane. If Macbeth's going to do the deed, he has to do it *now*.

I said previously that as you develop your play, identify the story, figure out where to start it, and create the pressures for action, the themes and ideas of your play would emerge. Actually, the details of the story and the thematic concerns reinforce each other; the point of the play is inextricably bound up with the selections you make regarding the details.

As we saw earlier, "Good Idea, If It Works" could move in any of several very different directions, depending on the story the author wants to focus on and what the author thinks is important. In other words, the story comes from the author's sense of values. An author must be willing to examine the most compelling concerns, for without a sense of values a plot is merely a mechanical contrivance and characters merely mechanisms for action.

Often the great themes of a play emerge in details that seem almost incidental to the action of the play. *Orphans* features a shadowy character who seems to have gangland connections. He comes into the house of two underprivileged orphans and changes their lives. He helps them procure jobs, fine food, and natty clothes. More importantly, he provides emotional support by giving them "an encouraging hug," which becomes a central visual motif of the play. When the man dies, the orphans try to place the dead man's arms around their shoulders.

During his stay with the boys, the visitor sings snatches from a particular old song, "If I had the wings of an angel." Although the man sings only a few lines, some of the lyrics of that song refer to poor sinners being enfolded by those angelic wings. In dialogue, too, the theme is expressed. After one of the young orphans is given a map of his neighborhood in Philadelphia—not accidentally the "City of Brotherly Love"—the boy declares, "I know where I am now." Indeed, this "lost boy" has now found himself.

The details of this play might have been different. The stranger could give the boys a pat on the shoulder rather than a hug, and he could sing another song. The orphan could speak other words. The setting could have been somewhere other than Philadelphia. But Kessler has chosen to dramatize the theme of encouragement through the details he has selected.

The following exercise is designed to explore concerns that are important to you. It is also an excellent device to get you started writing when you don't particularly feel the urge to write. The credit for devising this exercise, which is called Spontaneous Composition, goes to Jon Sedlak, a playwright who introduced it to me when he visited my playwriting class.

• • • **EXERCISE 8**
Write a monologue, which is a relatively lengthy speech by one person or character. You should not consider meaning, structure, grammar, punctuation, or even character. Rather, you should simply sit down and start writing about your own thoughts and feelings. The finished product should be about a

paragraph or page long. You should repeat the process two or three times, thus generating two or three different monologues.

This exercise, by itself, is not designed to produce a play. It can, however, accomplish several objectives. First, it will help you overcome the hesitation of starting something. It will allow you to write without the pressure that what you write has to be "good." These monologues are not finished products but rather stepping stones. Although they are not designed for formal production or evaluation, you should look at them carefully to see if they have potential for drama and character.

That potential leads to a second objective of the exercise. In many instances a monologue will be generated that can be used in some way in the development of a play. Perhaps a theme or emotion will emerge. Perhaps a certain phrase or group of words will suggest dialogue or character.

Yet a third reason for Spontaneous Composition is that if you are unhampered by the conventional constraints of character, plot, and dialogue, you will often discover a depth of feeling previously unknown about a subject. You may find an interest of which you were unaware. In short, this exercise will help you become aware of subconscious feelings and concerns.

In the following monologue, written by one of my students, I have not edited punctuation in order to give a clear picture of the product.

EXAMPLE

ARTISTIC CRAP
by John Thommasson

I heard him talking about it. He said Wow, it's
great. I never read anything like it. It
surrounds my world. It makes me realize how
things are. So I read it. I went out and plunked
down my hard-earned cash for a copy of the worst
thing I ever read. Artistic crap. He dances
around the subject without ever touching the
ground. It's meaningless. Or am I the one at
fault? Am I one more of the mindless millions,
destined forever to look at a thing of beauty and
say, Huh? What is it? What does it do? Why is it
done that way? It happens all the time. Maybe I'm
looking from the wrong angle. Or maybe I'm not
looking. But it was awful. All they do is talk
inane subjects that fall flat to the ears. The
sound of nothing on every page. But there must be
something there. Maybe you just have to learn to

```
enjoy it. Like the kids brought up on opera. I
don't understand that either. Two hours of
boredom listening to an unintelligible song. More
artistic crap. It's all over. It's all over.
```

It is easy to see from that example the strength of feeling generated through this assignment. It's also easy to see how parts of the monologue could spark other writing. The image of "The sound of nothing on every page," for instance, is worth exploring.

Another element that emerges as the play develops is style, which means the manner in which the play is done. Few authors set out to write a play in a particular style. Rather, the style of the play evolves from the material itself. In *Orphans* Kessler wanted the realistic trappings of a seedy Philadelphia apartment. In *Our Town* Thornton Wilder wanted to show us that we should, for our few hours on the stage of life, try to see and appreciate everything around us. To achieve his point, Wilder stripped the stage of scenery and props so that the audience would be forced to imagine all the shapes, colors, textures, sounds, and tastes of life with which Wilder stuffed the play. The style of "The Donut Shop" in Chapter 6 suggests a real world, while "Bad Luck" in Chapter 7, despite its realistic surface, avoids the details of an actual environment.

In line with that, I offer a warning regarding overt abstraction and abstract characters. It is easy to see good and evil in the world, and inexperienced writers are tempted to put those qualities on the stage as characters. In order to show a young man in conflict, an author may give us "Angel" and "Devil" presenting their cases in the boy's ears. Or "Good" and "Evil" will put in appearances. Seldom does that kind of abstraction work very well. Just as the essence of the character is defined in a name, such characters usually come across as predictable, one-dimensional voices. They lack complexity and the human qualities that most attract our attention.

Such abstractions are generally more successful when conceived in human terms. In the *Oh, God* series of movies, for instance, George Burns is endearing because of his idiosyncratic human foibles. The stranger in *Orphans* is a sort of angelic fairy godfather who transforms the boys' clothes and environment just as surely as Cinderella's fairy godmother transformed objects with her wand. But we care about the visitor because of his human interactions with the boys.

As you develop your play, you may find yourself running into some very mundane problems. "I'd really like to have a scene where John finds Barbara deceiving him with another man, but why would they get together if they've just had a fight?" If you need two characters in the same room at the same time for their climactic confrontation, you must solve the problem of how to get them there.

Playwrights must be practical problem solvers. They must be able to put all the pieces of the puzzle together. And just as more pieces usually make a jigsaw puzzle more challenging, a longer play makes the playwright's task more challenging.

This problem-solving facet of the playwright's job is a much underrated talent. Every time characters enter or exit, they must have a reason for coming or going—a reason that must be plausible. A beginning playwright might explain that a character visiting a friend "just drops by." Well, if friends don't do that in real life, they shouldn't do it in plays. Maybe the visitor is lonely and wants attention. Maybe the visitor is delivering a package or is just plain bored and wants the other character to go somewhere or do something to help pass time. Even that is a reason.

I was looking recently at Robert Bolt's *A Man for All Seasons*, a dramatization of Sir Thomas More's conflict with King Henry VIII over the King's marriage to Anne Boleyn. At one point (Act II, scene 7), More is in jail and his wife, daughter, and son-in-law visit him. They are permitted to see him only so that his daughter Margaret can try to persuade More to sign an oath of allegiance to the King.

The three are ushered into the cell area by a jailor, who lets More out of his cell for the visit. Greetings are exchanged, and the family give More some small gifts they've brought. Then we get to the heart of the scene. His daughter, Margaret, tries to persuade More to sign the oath. The real point of their good-natured argument is to display a loving relationship between the father and the daughter. She understands why he cannot sign. Even as she urges him to do it, she realizes that she would think less of him if he did.

The playwright exerts pressure on the scene. The jailor announces that the visit can last only two more minutes. More sends his son-in-law to occupy the jailor, and then he tells his wife and daughter to leave the country. He persists until they accede.

More's wife, Alice, is angry with her husband. She doesn't understand why his conscience is more important to him than his family. More tries to placate her and convince her, but she will be neither placated nor convinced. Finally, frustrated by his inability to make his wife understand, More breaks down. At that point Alice rushes to him and offers her support, not because she understands, but because she realizes her husband needs her.

The main transactions of the scene are done. The relationships between More and his daughter and More and his wife are finalized, so the scene can end. But Bolt does not end it gently. Disputes erupt as the jailor forces the family to leave.

There are 118 speeches in this scene. Fifty-three speeches, which is 45 percent of the scene, are devoted to the entrance, the greeting, the gifts, the "two-minute" warning, and the exits. All of those items are purely mechanical elements constructed by Bolt to surround and support the two crucial human relationships between More and his daughter and More and his wife, which constitute the core of the scene.

That is not unusual. In fact, Bolt probably invests his scenes with more substance than most playwrights. Good playwrights like Bolt solve the technical problems, and they are willing to take time to do it. They put the pieces together seamlessly. Good playwrights construct effective transitions to move from one segment to another. Good playwrights even use the mechanical elements to enhance the effectiveness of the core elements. Bolt doesn't waste 45 percent of the scene. He uses the conflicts with the jailor, for example, to heighten the tension of the scene and turn a simple exit into an emotion-wrenching departure.

The following exercise is designed to test your problem-solving abilities. It is structured to be used with a group of people, in a class or at a workshop, but I have included instructions to allow individuals to do the exercise.

• • • EXERCISE 9

Write a description of two characters. Each description should be only a sentence long. Write one sentence each about a place and an object. There does not need to be a connection between any of the four items described. The descriptions will be used to develop a scene. Bring your descriptions to class and exchange them with another student. Write a scene of about three or four pages using the characters, place, and object you have been given.

This exercise is in many important ways a reverse of the Spontaneous Composition assignment. Where that exercise sought to probe impulse and depth of feeling outside the conventional restraints, this exercise is pure problem-solving within the confines of a given location, given characters, and a given prop.

Within those defined limitations, the assignment is as much an exercise in creativity as the previous one. The student must solve the basic problem—What are those two characters doing in this place with that thing?—in a plausible, dramatic fashion.

Another strategic benefit of this exercise is that it increases a writer's awareness of environment and props. As soon as characters are placed within a locale, their behavior and therefore their actions are affected. And actions often involve doing something with objects. From the first silent scenes in this book—remember the woman with the Christmas tree and the ornaments?—to famous plays such as Laura with her glass menagerie, props help to focus the action. They play a vital and much underrated role in good drama.

Writers often go in one direction or another with this exercise. Some just throw up their hands and settle for anything, but others create truly inventive solutions to the problem. That, of course, is the point: to address those problem-solving skills that every play requires.

In the following exercise one of my students, Tim Carlin, was given a 38-year-old father, a teenaged daughter, a library, and a map.

EXAMPLE

> THE SETTING is the library of a house. JOHN, a large man of 38, sits at a desk. The desk top is covered with maps. One map is open on top of the others. JOHN studies this map, occasionally making marks on it with a felt-tipped pen. The walls of the room are covered with pictures of truckers and cowboys. There are many books on shelves around the room. A huge plush chair sits in the corner of the room.
>
> CATHY, a bright-eyed 16-year-old, RUSHES INTO THE ROOM filled with excitement and energy.

CATHY

Daddy! Guess what! I've been appointed chairman of the float committee. Well, chairperson, I guess. Anyway, I've got the best idea!

JOHN

That's great, honey.

CATHY

Listen to this. We're playing Kings High School, right? So we make a float saying 'Dethrone the Kings!' Then we have one of our guys dumping a dummy out of the throne! Get it?

JOHN

Sounds like a good idea. I could probably get a flatbed rig for the float if you want.

CATHY

I was hoping you'd say that. Thanks, Dad.
   (She kisses him.)
I gotta call Susie and go over what we need with her.
   (She goes to the door, then stops.)
Oh, I almost forgot. Can we use that big chair for the throne?

JOHN

Uh, oh. I just told your mother she could have it at her place.

                              CATHY
Aw, Dad. We need the chair!

                              JOHN
Well I'm sorry, but I already said she could have it.

                              CATHY
Couldn't you just give it to her after Homecoming?

                              JOHN
I won't be making a run out there for the rest of the
year. She'll want it before Christmas.

                              CATHY
Maybe you could explain to her that we need. . . . Oh,
she wouldn't care.

                              JOHN
You think she'll want you to use it for a Homecoming
float?

                              CATHY
Well, what am I going to tell Susie? I told everybody I
could get the chair.

                              JOHN
I'm sorry Cathy.

                              CATHY
No you're not. You want to see her. You don't care
about me. She's the one who left, remember? Not me!

                              JOHN
Cathy!

                    (CATHY slams the door ON HER WAY OUT. JOHN
                    walks to the desk and looks at the map a
                    long time. He slowly folds it up and drops
                    it in the trash can.)

                            THE END

The author managed to generate a strong relationship between the father
and the daughter in a very brief space. He also managed to create a gi-
gantic mood swing from the girl's first joyful entry to her frustrated exit.
He was less successful in his use of the map, which seems somewhat extraneous, and of the library, which provides an incongruous backdrop for
this scene.

     Obviously this exercise is designed for a group rather than for an
individual. However, to assist you in using it for yourself, I've included
six sets of character descriptions, places, and objects with which you can
create scenes of your own to test your own problem-solving talents.

Characters	Object	Place
An old man in faded denim overalls and a T-shirt	A ring of keys	The backyard deck of an upper-middle-income home
A minister with a clerical collar		
A youth in the uniform of a fast-food chain	A can of bug spray	A playground
A woman, 26, obviously pregnant		
A teenage baby sitter	A fishing pole	A hospital lobby
An overweight, middle-aged woman with dyed blonde hair		
An elderly man in a tuxedo	A guitar	A bus terminal
A dowdy woman of 45		
An 80-year-old stroke victim	A mask	A public library
An energetic eight-year-old		
A girl of 18 in her pajamas	An envelope containing photographs	A cabin
A thin young man in jeans with a baseball hat turned backwards on his head		

Now you're fully equipped to put together your own play, so let's proceed to Exercise 10.

• • • **EXERCISE 10**
Write a one-act play about 20 to 40 minutes in length. It can be an extension of a previous scene or an entirely new creation. Use correct playwriting format.

Before you begin your own piece, you might want to read this play, which was written as a final project by one of my students. Because of space limitations, I have selected a relatively short one-act play.

EXAMPLE

```
            GET BACK IN LINE
            by William Croxton

THE SETTING is the front of a ticket office
window. TERRY and JULIE are sitting on the
ground in front of the window. A poster
above their heads shows an androgynous
figure with lots of hair, minimal clothing,
```

and arms outstretched to the sky. The words
"Tickets on Sale" appear across the poster.
TERRY and JULIE are dressed in winter
clothing and have sleeping bags. TERRY is in
his late twenties and is wearing a concert
T-shirt over the top of his coat. He is very
excited, almost wired, and jumps up and down
from his seated position to look at the rest
of the crowd, which is indicated merely by
cutout silhouettes. JULIE, almost 30, sits
grading papers and doing her best not to pay
too much attention to TERRY.

                    TERRY
        (He finishes the liquid in a thermos bottle,
        then gets up and says, overdramatically.)
The electricity fills the scene!

                    JULIE
I'm freezing. Give me some more coffee.

                    TERRY
This is too good to be true. First in line!

                    JULIE
Terry? More coffee please.

            (She pushes her cup toward him.)

                    TERRY
I wonder if he knows we're here? Do you think so?
            (He notices the cup she's holding up.)
Coffee's all gone.

                    JULIE
What? Are you kidding?

                    TERRY
Hey, check the thermos if you don't believe me.

                    JULIE
Terry, this is not a happy face. We have six more hours
before this stupid ticket window opens. I'm frozen. And
we're out of coffee. Are you listening to me?

                    TERRY
Something like that.
        (Checking watch)
Actually six hours and seven minutes. But don't worry.
Wacky Waldo said on the radio that he and the rest of

the Morning Zoo are bringing breakfast to everyone in line.

        JULIE

Great.

        TERRY

Besides, who needs it? Can't you feel the energy?

        JULIE

I'm sure you can. You drank all the coffee.

        TERRY

We'll be fine.

        JULIE

Wacky Waldo--I can't believe I just said that--isn't going to be here until at least six. What do we do for two hours?

        TERRY

Up for a round of TV trivia?

        JULIE

No thanks.

        TERRY

I don't know. Where are the magazines we packed?

        JULIE

They're not here.

        TERRY

What? Well, where. . . .

        JULIE

I didn't bring them.

        TERRY

You didn't bring my magazines?

        JULIE
     (Assertiveness training)
No, I didn't. I thought as long as we were going to stay outside all night . . .
     (Losing it)
. . . that we could . . . well . . . use the time to, ah, . . . talk.
     (Pause)
Can't we talk anymore?

        TERRY

What do you want to talk about?

> JULIE

Us.

> TERRY

Us?

> JULIE

Yes. Us.

> TERRY

You and me. Okay. We'll talk about us. Terry and Julie. This is Terry and Julie talking about Terry and Julie. Well. I think we're doing just fine. I mean, look at us. We're first in line to see just about the biggest rock concert in the world, and now since that's covered I'd like, really like to read my magazines if they were here.

> JULIE

Come on, Terry. That's not what I'm talking about. But, kind of along those lines, don't you feel a little . . . foolish?

> TERRY

Foolish?

> JULIE

Yeah. A little? Look around. We're sitting in the middle of a crowd of 15-year-olds to get tickets to see some over-hyped, no-talent, teen icon!

> TERRY
> (Putting his hand over her mouth)

Okay, Julie, first of all, we're in front! Second, this "over-hyped, no-talent, teen icon" is a phenomenon. A very interesting and intriguing study in human behavior.

> (He slowly removes his hand, waiting for a response.)

> JULIE

And let's not forget your scholarly interest in prepubescent girls!

> TERRY
> (Quickly putting his hand back over JULIE's mouth)

Thirdly, may I remind you, that you, too, are in line.

> (He again removes hand cautiously.)

                         JULIE
Don't remind me.

                         TERRY
Now just sit back and stay awake. If we don't, somebody
might sneak in front of us.

                         JULIE
So what?

                         TERRY
So what? The TV and radio guys never interview the
second people who buy tickets.

                         JULIE
Believe me, I'll stay awake. I'm too cold to sleep.
                (She resumes her paperwork. Finally, after
                TERRY sits back down, JULIE inches closer to
                him pressing herself against him. TERRY
                remains oblivious to her advances.)
What is it? Please tell me. Am I not pretty anymore?
Are you doing this to drive me nuts? Do you want to be
with someone else?

                         TERRY
Hey, slow down, willya?

                         JULIE
I just want to know what's going on.

                         TERRY
Nothing.

                         JULIE
I feel like you're dumping me for a poster person of
indeterminate sex.

                         TERRY
You don't get it, do you?

                         JULIE
No. I guess not.

                         TERRY
You're a high school teacher. You should know these
things.

                         JULIE
What things?

                         TERRY
Let me see this stuff.
                (He reaches for her papers.)
This is what? Algebra?

                        JULIE

Geometry.

                        TERRY

Fine. Geometry. Now what does the average kid care, or
for that matter what does he need to care about this
stuff?

                        JULIE
                (Grabbing papers back)
Absolutely nothing. You're right. Why go to school at
all?

                        TERRY

I'm not saying that. But . . . like . . . how many days
can you say that you've lived that have been . . . well
. . . drastically changed because of geometry?

                        JULIE

God, this is making me feel so much better.

                        TERRY

You still don't get it.

                        JULIE

No, I don't.

                        TERRY

Julie, you could learn so much from these kids.

                        JULIE

I could?

                        TERRY

Yes! I mean . . . this singer has more power than the
president. The fans at his disposal would translate
into about a billion voters. . . .

                        JULIE

All eagerly awaiting their first pubic hair.

                        TERRY

Funny.

                        JULIE

Okay. I'm sorry. But let's keep some perspective here.
How, for instance, would your favorite rock singer of
the moment solve the world's problems? How would he
. . . say . . . prevent financial ruin in the economy?

                        TERRY
                (Brightly)
A telethon! Yeah! A huge telethon. Everybody would
come.

                    JULIE
All the stars descending from show-biz mountain.

                    TERRY
It's happened before.

                    JULIE
Do you realize how utterly ridiculous you sound? You
throw out all these really easy solutions to all the
world's problems . . .
          (TERRY pulls out a Walkman and puts it on.)
Oh, no you don't! You're not pulling that one again.
Take it off. Take it off right now. I'm warning you.
I'll leave.
          (JULIE tries unsuccessfully to catch his eye
          or get his attention.)
I'll leave. That's it. I'm leaving.
          (Very loud, near his head)
I'm leaving.

          (She gets up and starts to gather her
          things.)

                    TERRY
Go on, but you'll be sorry you missed it.

                    JULIE
          (In the midst of collecting her belongings)
I can't leave. Dammit. I want to leave. I want to be
home in my own bed, but I can't.

                    TERRY
What?

                    JULIE
          (Yelling)
I can't leave!

                    TERRY
Wait a second until this song's over.
          (JULIE yanks the headphones off.)
Ow! What'd you do that for?

                    JULIE
I can't go home.

                    TERRY
Why not?

                    JULIE
We hitchhiked, remember?

                    TERRY

Oh yeah.

                    JULIE

Something about getting more of a feel for the night.

                    TERRY

Right.

                    JULIE

Besides, we have to talk.

                    TERRY

I thought we did.

                    JULIE
            (Determined to say what she wants to say)
Terry, you have to grow out of this phase.

                    TERRY

What phase?

                    JULIE

This extended adolescence of yours. This study of all
things MTV. You're 28 years old, Terry.

                    TERRY

And?

                    JULIE

And. . . .

                    TERRY

This coming from the 29-year-old lips of "We're not
living together, we're learning together"!

                    JULIE

Nice shot.

                    TERRY

I'm not trying to be mean. I just think you're
overreacting.
            (Short pause)
It's because you're older than me.

                    JULIE

What? I'm one year older, for Pete's sake.

                    TERRY

It's a crucial year.

                    JULIE

Seven months. Are you saying I don't relate to the
youth culture as well as you do because of seven
months?

                    TERRY
It's just a hunch, but maybe. There have been studies.

                    JULIE
In one of your music magazines, no doubt. I can see the
headline now: "Un-hip 29-year-old found." You're losing
your mind.

                    TERRY
No. You're losing your ability to look unbiased at the
situation. But don't worry. That's what I'm here for.

                    JULIE
Hurray.

                    TERRY
Look. I know you turn 30 in two months. I think that's
the problem.

                    JULIE
Two and a half months. But that has nothing to do with
any of this. You have to grow up.

                    TERRY
Yep. That's what it is. Just don't worry about it so
much.

                    JULIE
Terry, you're acting like a child.

                    TERRY
I'm absorbing my culture. You could use a dose
sometime.

                    JULIE
I'm up to my neck in it.

                    TERRY
So don't waste it. Use it. Can't you see that life
becomes so much more fun that way? There are all these
heroes and dreams and fantasies. . . .

                    JULIE
And they all fit neatly inside a 27-inch TV screen.

                    TERRY
Exactly! It's in all of us!

                    JULIE
Well, it's not inside me. That's the most depressing
thing I've ever heard.

                    TERRY
Only on the inside. On the outside it sets up a warmly

content society unable to voice its frustrations because they're afraid they'll miss something.

                    JULIE
Did you read that somewhere?

                    TERRY
No. I think I made it up.

                    JULIE
Don't you see what a shallow existence that is?

                    TERRY
What's shallow about a good feeling?

                    JULIE
Mere contentment isn't a real feeling. There aren't any real feelings in rock star worship. It's all so de-humanizing. We need the warmth of other real human beings.

                    TERRY
Then how come these real, warm human beings do awful things to each other?

                    JULIE
I'm not saying everything is great. But at least it has some emotion to it. I mean, if there's a single speck of humanity somewhere, you can't turn your back on it.

                    TERRY
Wait a minute. I think I saw a piece on "Entertainment Tonight" on that stuff.

                    JULIE
Oh, Terry.

                    TERRY
No, really. Yeah, it was some sort of rally in Hollywood to bring the human race closer together. All the biggies were there. They held hands and sang a song and afterwards everybody got awards for it.

                    JULIE
Yeah, well, you see? We aren't alone. Doesn't it make the world seem like a better place to live?

                    TERRY
Oh yeah. I mean, I guess I've always felt like that. You know. That human beings should always--
                    (Suddenly pointing)
Hey! Look! It's Wacky Waldo!

                         JULIE
                  (Jumps over to see)
Where?

                         TERRY
There. See him?

                         JULIE
Which one's him? I don't see him.

                         TERRY
It's not him.

                         JULIE
Huh?

                         TERRY
I was kidding.

                         JULIE
What?

                         TERRY
Look at yourself. Doesn't that tell you something? You
wanted to see Wacky Waldo.

                         JULIE
You jerk.

                         TERRY
You <u>leapt over me</u> to see Wacky Waldo, and he's just an
ordinary deejay. . . .

                         JULIE
                  (Overlapping)
Look, creep. . . .

                         TERRY
You wanted to see him. You know you did.

                         JULIE
I just wanted the coffee!

                         TERRY
It's okay, Julie. I just wanted to help you let your
instincts out.

                         JULIE
Terry, I didn't really care. . . .

                         TERRY
Now that everything is out, we can have that talk you
wanted to have.

                              JULIE
You are such a. . . .

                              TERRY
What do you want to talk about? Top Ten?

                              JULIE
I don't believe you.

                              TERRY
Grammys?

                              JULIE
                   (Trying for controlled and mature)
I am not going to get mad. Anger never helped a
situation. I'm going to sort it out like laundry. No
anger. I'm just going to get my things and leave.

           (She gathers her things.)

                              TERRY
What do you mean, leave? You're finally out. Why would
you leave?

                              JULIE
If you are unable to comprehend the severity of the
last few minutes then it only confirms the fact that
you and I have become completely incompatible.

                              TERRY
Because you like Wacky Waldo more than I do?

                              JULIE
Terry, we have arrived at a very serious impasse in our
relationship. I wish you would treat it with a little
more concern. What am I saying? You aren't even living
in the real world anymore.

                              TERRY
What could be more real than this?

                              JULIE
Us. Or what used to be us. We're more real. Whatever
became of that?

                              TERRY
Julie, I think you're looking at the wrong things.

                              JULIE
I'm not looking at the wrong things! Where do we go
from here? We can't stay here in this one spot any
longer. We have to move on. Or at least I have to.

                    TERRY
Julie, this is my home. This is where I belong.
Watching all of this.

                    JULIE
You can't watch for the rest of your life.

                    TERRY
Why not?

                    JULIE
          (Speaking quickly, the facade breaking)
'Cause it's not fair. It's against the rules. You can't
just stop and watch. You have to grow up.

                    TERRY
Whoa! Whoa!

                    JULIE
You'll end up losing everything to something bigger.
It's too easy to succumb. . . . (resigned) and it's too
hard to try.
          (Pause. They both lean back against the
          wall.)

                    TERRY
So what happens now?

                    JULIE
I don't know. Nobody ever knows.

                    TERRY
We really do look a little silly here. Come on. I'll
call a cab.

                    JULIE
Are you kidding? We're first in line.

                    THE END

     The focus, of course, is the relationship. Julie wants it to move to
another step, and for that to happen she thinks Terry will need to grow
up. Although there is not a great deal of physical action and the play is
about "talking," the author is so clear on what the characters want that
the piece never becomes just dialogue.
     Julie has planned this carefully. She intentionally did not pack
Terry's magazines. She knows that once they are in line, Terry can't run
off anywhere. And Julie can't leave either. Pressure comes from that
situation.
     I particularly like the resolution. Julie, apparently the more con-

trolled and ordered of the two, seems less able to express herself fully. Her arguments are easily beaten back by Terry's insistence that pop culture exerts enormous influence on society. Yet, at the end, when Julie finally breaks into almost incoherent expressions, Terry's real concern for her comes through. He's even willing to give up his place in line. Julie senses she's won something. For her, his willingness to leave is enough, and so she agrees to stay. One suspects, however, that they won't be in lines like this again very often.

Once you've written your own play, you will want to examine it as objectively as you can, just as I have done with the scenes in this book. When you're ready for a self-evaluation, begin by asking yourself these questions:

—Who is the main character? Who is the play about?

—What does this character want?

—What are the obstacles to the desire of this main character?

—Does this character undergo any changes in the course of the play?

—What is the most important moment—the climax—of the play?

—Why should we, the audience, care about what happens to this character?

—Can you summarize the action of your play concisely, as in the following format:

> This play is about a [PERSON: for instance, "a woman," "a man," "a boy," "a young girl"] who [ACTION PHRASE OF A FEW WORDS: for instance, "wants to be king," "seeks revenge for his father's murder," "pursues the truth of an unsolved murder"].

If we apply these questions to "Get Back in Line," we see that Julie is the main character, and she wants a more mature relationship with Terry. Her obstacles include Terry's fascination with childish pleasures and his reluctance to accept responsibility. The key moment occurs when Julie finally gives up, and Terry acknowledges his own silliness. With that acknowledgment, Julie achieves at least part of her goal. We care about Julie not just because she tries hard, but because we all have relationships that we would like to strengthen but are reluctant to confront for fear of destroying them instead. We could summarize the action as follows: This is a play about a young woman who wants to deepen her relationship with her boyfriend.

If you have trouble answering those questions about your play, then you need to focus your action more carefully as you work on your second draft. And then it's on to your finished play. You've identified your story. You know where to start it. You understand what you want to express about the story. You've placed the characters in conflict, and you've put them under pressure. You've included significant details. You've always kept in mind the human qualities of your characters. And out of that grows your play—complete, powerful, intriguing, sensitive, witty. Your play will astonish the audience. It will entertain them, and it will move them to laughter and tears.

WHAT TO DO WITH YOUR PLAY

And then what? A play isn't meant just for you. It's meant for other people. So the next question is, What do you do with your play once you've written it?

First, a few things *not* to do. There is little point in submitting your play to a publisher or to a play-leasing company. A publisher such as Hill and Wang, Grove Press, or Random House is only interested in publishing a play for its literary value *after* it has already achieved success as a theatrical production. Similarly the play-leasing companies such as Samuel French or Dramatists Play Service are normally interested in putting out a production version of a script and arranging for production rights only *after* the play has received a successful run.

So the question really is, How do you go about getting your play produced? First, if there are school, community, or regional theaters in your area, you should consider taking an active part. Attend the plays. Work on the productions if you can. Get to meet the other people who are involved. Your play has a much greater chance to be read or produced by people you know than by strangers.

Second, there are many amateur and professional theaters that actively solicit new scripts. The best way to find out about these theaters and the kinds of scripts they want is to look at a book called *The Playwright's Companion*. Compiled by Mollie Ann Merserve and published annually by Feedback Theatre Books, the book identifies theaters that are willing to look at original scripts. The theaters range from the most prestigious New York, Chicago, and West Coast companies to professional regional theaters to college and university theaters to semiprofessional, amateur, and community theaters. A recent edition listed over 500 theaters but not every one of them will necessarily produce original scripts every year. And if they do, the playwright is likely to be an already established author.

A third approach to production is to enter your play in appropriate contests. You can be assured of a careful reading, and in most cases an actual winner or winners are named. A recent edition of *The Playwright's Companion* identified over 150 play contests and listed various criteria such as deadlines and contact persons. In most cases prizes ranged from $100 to $5,000. In almost all cases the winning scripts are assured of production. Theaters that produce prize-winning plays often bring the playwright in for the production and sometimes for the rehearsal process as well.

Requirements for the contests vary. Some are specifically for young writers, some accept only one-act plays, others want only full-length plays. Some contests exclude musicals while others are *only* for musicals. There are several contests for children's scripts.

In many contests entrants must be from a particular state, area, or region. In a few the author must be of a specific race or gender. Contests quite often include rules about the sets or numbers of characters that should be used. Most theaters are looking for scripts with minimal scenic requirements, such as one set and a handful of characters. Theaters today are looking especially for scripts with good female roles.

There are contests for translations of plays from Spanish, Hebrew, and Scandinavian languages into English. Whenever authors are dealing with a translation or an adaptation, they must be careful to secure the appropriate rights to the original material.

Sometimes requirements for subject matter are specified. There are contests that seek scripts about animal rights, Afro-American experiences, Jewish values, women's issues, Hispanic-American concerns, and rural America. One contest is for plays about Hawaii. Another requires that entries focus on events within the past 200 years, deal with issues raised by the Bill of Rights or by the Fourteenth and Fifteenth Amendments, *and* be based on records held by the National Archives or by a presidential library!

If you decide to submit one of your plays to a contest, you should follow the playwriting format illustrated in the scenes in this book, with margins of at least one inch on three sides and one-and-one-half inches on the left to allow for binding. For a one-act play you would use consecutive page numbering in the upper right hand corner. If the play has more than one act, pages are numbered by act (for example, II-12). Acts are given capital Roman numerals, and scenes are written with small Roman numbers (for example, II-iii-17).

A title page should include the title of the play, a brief designation such as "a comedy in one act," and the author's name. These should be just above the center of the page. At the bottom right is the playwright's address, and in the bottom left the copyright notification. Authors may copyright unpublished and unproduced plays by writing to the Copyright Office in Washington, D. C., and requesting copyright forms.

A second page includes the cast of characters. Each character should be listed, followed by a brief description. Listings for time and place for each act and scene should also be included on this page.

One problem with submitting plays to theaters or to contests is the lack of feedback. Because they receive so many scripts, theaters and contests seldom send playwrights any comments on their work. Playwriting seminars or writing workshops can often provide valuable responses.

Getting a play produced can be a long and arduous process. Sometimes the timeliness of a current issue evaporates before the play is ever performed. Frequently plays must be rewritten again and again to satisfy the needs of producing organizations. Despite all the hurdles, an undeniable and unforgettable magic occurs when other artists—actors, directors, and designers—take your script and add their talents to it to bring forth a living, artistic creation.

Appendix

• • • • • • • •

Classroom Procedures

Some readers of this book may want to use these exercises in a classroom setting. The sequence of exercises I've outlined has been used successfully by instructors at the Virginia Summer program for the Gifted and Talented as well as by English and drama teachers who wanted to develop a unit on playwriting for their classes.

Because I believe that *how* something is done shares importance with *what* is done, I've suggested some ways of proceeding. Because I know that teachers and their classes vary, I offer these suggestions as just that: suggestions.

1. The student should read his or her own work to the group.

There are a variety of ways to disseminate to the rest of the class a scene written by a student. Some instructors like to make copies of the piece, pass them out, and have everyone read the scenes outside of class time. That may be necessary if there is insufficient class time, but you lose the sound of the scene, which I think is a large sacrifice.

Many instructors like to assign the parts in a scene to the students in class. That way, different people read different parts, more students are actively involved, and the author can hear what it sounds like with other people acting the words. This technique also offers the added dimension of what the "performers" reading the role contribute to it by their delivery of lines.

I like to have the student's work read aloud, and I like to have students read their own pieces. I want to get as close to the author's conception as possible. Therefore, I want to hear a line the way the author wants it read. Although another member of the class may contribute a different quality to the line, at this early stage I don't particularly want that contribution. I want to make the connection from the author's mind to the work we hear as direct as possible.

This procedure can result in some confusion about which character is speaking, and I often have to remind a student to slow down and to read the name of the character speaking before reading the line. Nevertheless, I still prefer this method.

Finally, by reading their pieces themselves, students must take responsibility for what they're writing. In professional writing situations authors must be able to "pitch" their material; they must be their own best salespeople. Similarly I believe student writers should give us the best readings they can muster.

> 2. Except for reading the piece, the writer should refrain from all editorial comment until all remarks are concluded.

I've often run into students who are great talkers and want to *tell* us about their scenes. That must be avoided. No explaining before the scene starts. No justifying after it's over. The dramatic material must, like the cheese, stand alone. The audience should be responding to the scene and only to the scene, not to the author's commentary on the scene.

> 3. As many students as is practical should comment on each work that is read.

I'll divide a group of more than a dozen students in half, and ask half the people to comment on a scene. With fewer than that, everyone comments on every piece. This practice generates a climate in which each person exercises analytical skills on every play, and each feels somehow a part of every other student's play. Instructors may also find it helpful for students to keep a notebook or journal of their comments, questions, and reactions to works that are presented.

Every student will not be an exceptional playwright. But every student can gain an understanding of the process and the problems of writing plays. Every student can attain an appreciation for the difficulties of structuring plot, developing character, instigating conflict, and devising dialogue. Students learn from hearing, analyzing, and commenting as well as from their own writing.

For those reasons, I regard the commenting time as essential. At first students may be reluctant to critique the work of others. They may feel they don't know how to do it, what to look for, or what the teacher wants. They may need guidelines, such as "Start out by saying what you liked about the piece, what caught your attention or interested you. You can also mention things you didn't like or didn't understand." You will probably want to steer students away from rewriting other students' work, which they sometimes try to do by suggesting changes or indicating how they would solve a problem or make something better.

After everyone has had an opportunity to respond to a work, the author may make comments or answer questions.

Bibliography

Anderson, Maxwell. *The Essence of Tragedy*. New York: Russell and Russell, 1970.

Archer, William. *Play-making*. Boston: Small, Maynard and Co., 1912.

Armer, Alan A. *Writing the Screenplay*. Belmont, Calif.: Wadsworth, 1988.

Baker, George Pierce. *Dramatic Technique*. New York: De Capo Press, 1971.

Berman, Robert. *Fade In: The Screenwriting Process*. Studio City, Calif.: M. Wiese Film-Video, 1988.

Blacker, Irwin R. *The Elements of Screenwriting*. New York: Macmillan, 1988.

Brady, Ben. *The Keys to Writing for Television and Films*. Dubuque, Iowa: Kendall/Hunt, 1982.

Brady, Ben, and Lance Lee. *The Understructure for Writing for Film and Television*. Austin: University of Texas Press, 1988.

Brady, John, ed. *The Craft of the Screenwriter: Interviews with Six Celebrated Screenwriters*. New York: Simon & Schuster, 1982.

Bronfeld, Stewart. *Writing for Film and Television*. Englewood Cliffs, N.J.: Prentice-Hall, 1981.

Busfield, Roger M., Jr. *The Playwright's Art*. Westport, Conn.: Greenwood Press, 1971.

Cassady, Marshall. *Characters in Action: A Guide to Playwriting*. Lanham, Md.: University Press of America, 1984.

Catron Louis E. *Writing, Producing, and Selling Your Play*. Englewood Cliffs, N.J.: Prentice-Hall, 1984.

Chekhov, Michael. *To the Director and Playwright*. Compiled by Charles Leonard. Westport, Conn.: Greenwood Press, 1977.

Cohen Edward M. *Working on a New Play*. New York: Prentice-Hall Press, 1988.

Cole, Toby, ed. *Playwrights on Playwriting*. New York: Hill and Wang, 1961.

Dmytryk, Edward. *On Screen Writing*. Boston: Focal Press, 1985.

Edmonds, Robert. *Scriptwriting for the Audio-Visual Media*. New York: Teachers College Press, 1984.

Egri, Lajos. *How to Write a Play*. New York: Simon & Schuster, 1942.

———. *The Art of Dramatic Writing*. New York: Simon & Schuster, 1972.

Fann, Ernest L. *How to Write a Screenplay*. Los Angeles: David-Kristy, 1988.

Field, Syd. *Screenplay: The Foundations of Screenwriting*. New York: Dell, 1982.

———. *The Screenwriter's Workbook*. New York: Dell, 1984.

Finch, Robert. *How to Write a Play*. New York: Greenberg Press, 1948.

Frankel, Aaron. *Writing the Broadway Musical*. New York: Drama Book Specialists, 1977.

Freytag, Gustav. *The Technique of the Drama*. New York: B. Blom, 1968.

Funke, Lewis, ed. *Playwrights Talk About Writing*. Chicago: Dramatic Publishing Co., 1975.

Geller, Stephen. *Screenwriting: A Method*. New York: Bantam Books, 1984.

Gibson, William. *Shakespeare's Game*. New York: Atheneum, 1978.

Gillis, Joseph. *The Screenwriter's Guide*. New York: Zoetrope, 1987.

Giustini, Roland. *The Filmscript: A Writer's Guide*. Englewood Cliffs, N.J.: Prentice-Hall, 1980.

Goldman, William. *Adventures in the Screen Trade: A Personal View of Hollywood and Screenwriting*. New York: Warner Books, 1983.

Goodman, Evelyn. *Writing Television and Motion Picture Scripts That Sell*. Chicago: Contemporary Books, 1982.

Granville-Barker, Harley. *On Dramatic Method*. London: Sidgwick and Jackson, 1931.

Grebanier, Bernard. *Playwriting*. New York: Barnes & Noble, 1961.

Griffiths, Stuart. *How Plays Are Made*. Englewood Cliffs, N.J.: Prentice-Hall, 1984.

Haag, Judith H., and Hollis R. Cole, Jr. *The Complete Guide to Standard Script Formats*. Hollywood: CMC Publishing, 1980.

Hatton, Thomas J. *Playwriting for Amateurs*. Downer's Grove, Ill.: Meriwether Publishing, 1981.

Hauge, Michael. *Writing Screenplays That Sell*. New York: McGraw-Hill, 1988.

Herman, Lewis. *A Practical Manual of Screen Playwriting for Theater and Television Films*. Cleveland: World Publishing, 1966.

Hull, Raymond. *How to Write a Play*. Cincinnati: Writer's Digest Books, 1983.

———. *Profitable Playwriting*. New York: Funk & Wagnalls, 1968.

Josefsberg, Milt. *Comedy Writing for Television and Hollywood*. New York: Perennial Library 1987.

Kerr, Walter. *How Not to Write a Play*. New York: Simon & Schuster, 1955.

King, Viki. *How to Write a Movie in 21 Days*. New York: Harper & Row, 1988.

Kline, Peter. *Playwriting*. New York: R. Rosen Press, 1970.

Korty, Carol. *Writing Your Own Plays*. New York: Scribner, 1986.

Langer, Lawrence. *The Play's the Thing*. Boston: The Writer, Inc. 1960.

Lawson, John Howard. *Theory and Technique of Playwriting and Screenwriting*. New York: Garland Publishing, 1985.

Lee, Donna. *Magic Methods of Screenwriting*. Tarzana, Calif.: del oeste Press, 1978.

Longman, Stanley Vincent. *Composing Drama for Stage and Screen*. Boston: Allyn & Bacon, 1986.

Macgowan, Kenneth. *A Primer of Playwriting*. Westport, Conn.: Greenwood Press, 1981.

Maloney, Martin, and Paul Max Rubenstein. *Writing for the Media*. Englewood Cliffs, N.J.: Prentice-Hall, 1980.

Marion, Frances. *How to Write and Sell Film Stories*. New York: Garland Publishing, 1978.

Matthews, Brander, ed. *Papers on Playmaking*. Freeport, N.Y.: Books for Libraries Press, 1970.

Mayfield, William F. *Playwriting for Black Theatre*. Pittsburgh: William F. Mayfield, 1985.

Merserve, Mollie Ann, comp. *The Playwright's Companion: A Submission Guide to Theatres and Contests in the U.S.A.* New York: Feedback Theatre Books, published annually.

Miller, J. William. *Modern Playwrights at Work*. New York: Samuel French, 1968.

Miller, William C. *Screenwriting for Narrative Film and Television.* New York: Hastings House, 1980.

Muth, Marcia. *How to Write and Sell Your Plays.* Santa Fe, N.M.: Sunstone Press, 1974.

Nash, Constance, and Virginia Oakey. *The Screenwriter's Handbook,* New York: Barnes & Noble, 1978.

Nigli, Josefina. *New Pointers on Playwriting.* Boston: The Writer, Inc., 1967.

Nolan, Paul T. *Writing the One-Act Play for the Amateur Stage.* Denver: Pioneer Drama Service, 1977.

Norton, James H., and Francis Gretton. *Writing Incredibly Short Plays, Poems, and Stories.* New York: Harcourt Brace Jovanovich, 1972.

Packard, William. *The Art of the Playwright.* New York: Paragon House, 1987.

Pike, Frank, and Thomas G. Dunn. *The Playwright's Handbook.* New York: New American Library, 1985.

Polsky, Milton E. *You Can Write a Play.* New York: Rosen Publishing Group, 1983.

Polti, Georges. *The Thirty-Six Dramatic Situations.* Boston: The Writer, Inc., 1954.

Rilla, Wolf. *The Writer and the Screen.* New York: William Morrow, 1974.

Root, Wells. *Writing the Script.* New York: Holt, Rinehart, & Winston, 1980.

Rowe, Kenneth T. *Write That Play.* New York: Minerva Press, 1968.

Sautler, Carl. *How to Sell Your Screenplay.* New York: New Chapter Press, 1988.

Savran, David, ed. *In Their Own Words: Contemporary American Playwrights.* New York: Theatre Communications Group, 1988.

Seger, Linda. *Making a Good Script Great.* New York: Dodd, Mead, 1987.

Shanks, Bob. *The Primal Screen.* New York: Fawcett, 1987.

Smiley, Sam. *Playwriting: The Structure of Action.* Englewood Cliffs, N.J.: Prentice-Hall, 1971.

Straczynski, Michael J. *The Complete Book of Scriptwriting.* Cincinnati: Writer's Digest Books, 1964.

Swain, Dwight V. *Film Scriptwriting.* Boston: Focal Press, 1988.

Vale, Eugene. *The Technique of Screen and Television Writing.* New York: Simon & Schuster, 1986.

Van Druten, John. *Playwright at Work.* New York: Harper & Brothers, 1953.

Wager, Walter, ed. *The Playwrights Speak.* New York: Dell, 1968.

Walter, Richard. *Screenwriting.* New York: New American Library, 1988.

Weales, Gerald Clifford. *A Play and Its Parts.* New York: Basic Books, 1964.

Whitcomb, Cynthia. *Selling Your Screenplay.* New York: Crown Publishers, 1988.

Willis, Edgar E., and Camilee D'Arienzo. *Writing Scripts for Television, Radio, and Film.* New York: Holt, Rinehart, & Winston, 1981.

Wolff, Jurgen M., and Kerry Cox. *Successful Scriptwriting.* Cincinnati: Writer's Digest Books, 1988.

Index